FOREST GLEN ELEMENTARY SCHOOL
1935 DISCARDED NE
GREEN BAY, WI. 54313

EXPLORING SPRING

BOOKS BY SANDRA MARKLE

Exploring Winter
Exploring Summer
Exploring Spring
Science Mini-Mysteries
Power Up

EXPLORING SPRING

by Sandra Markle

*A Season of Science
Activities, Puzzles,
and Games*

FOREST GLEN ELEMENTARY SCHOOL
1935 CARDINAL LANE
GREEN BAY, WI. 54313

Atheneum   1990   New York

For Paige and Dort, good friends in all seasons

Copyright © 1990 by Sandra Markle

All rights reserved. No part of this book may be reproduced or transmitted in any form or by any means, electronic or mechanical, including photocopying, recording, or by any information storage and retrieval system, without permission in writing from the publisher.

Atheneum
Macmillan Publishing Company
866 Third Avenue, New York, NY 10022
Collier Macmillan Canada, Inc.
First Edition
Printed in the United States of America
10 9 8 7 6 5 4 3 2 1

Library of Congress Cataloging-in-Publication Data
Markle, Sandra.
Exploring spring / written and illustrated by Sandra Markle.
—1st ed. p. cm. Includes index.
Summary: A collection of springtime activities that include stories, observations of nature, handicraft, games, and puzzles.
ISBN 0-689-31341-1
1. Spring—Juvenile literature. [1. Spring. 2. Amusements.]
I. Title.
QH81.M2655 1990
574.5′43—dc19
89-394 CIP AC

CONTENTS

Is It Spring Yet?   1

1. SPRING FEVER   3
   Early Warning Signs · For the Record · Lion Hunting · Scrub It! · Clean Fun · Dump It · Reuse It · Spring Tonic · Berry, Berry Good · Big Hairy Deal · Let's Molt Again · Let's Play

2. FAMILY TIME   25
   Going Courting · Bowerbirds · Snowshoe Rabbits · Hummingbirds · Pigeons · Grizzly Bears · Siamese Fighting Fish · I Love How You Smell! · Falling in Love · No Trespassing! · Animal Nurseries · American Alligator · Skunks · Opossums · Deer · Become a Bird Landlord · How Hard Is It to Build a Robin's Nest? · Home Sweet Nest · Help a Bird Build a Nest · House for Dinner

3. HAPPY BIRTHDAY, BABY   41
   Having Babies · Laying an Egg · Eggsamination · A Naked Egg · Eggsperiments · Good Egg · Off to the Egg Races · Rare Eggs · Tough Stuff · Mermaid's Purse · Giving Birth · Mother's Milk · Fast Baby Food · An Orange Moo · What's the Baby's Name? · Growing Down · Nose Job · Taking Care of Baby · The Baby-sitting Business

4. BUDS AND BLOOMS   61
   From Flowers to Seeds · Flower Helpers · Smell Hunt · Ha-a-a-a Choo! · Wildflowers · Hepatica · Bloodroot · Longspur Violet · Red Trillium · Pink Lady's Slipper · Jack-in-the-Pulpit · Whistling

Dixie · Sap's Rising · Surprise Package · Blooming Trees · Pussy Willow · Horse Chestnut · Sugar Maple · Lettuce Eat · Pocket Salad · Crazy about Tulips

5. WILD AND WINDY   85
Why Is March Windy? · How Windy Is It? · Flights of Fancy · Kite Builder's Kit · Kite Flyer's Safety Tips · Play It Again, Wind · The Storm · Sounds Like Thunder · A Portable Roof · Give Me a Sign · How Rainy Is April?

6. STRICTLY SEASONAL   105
April Fool · Hiding · Not What It Seems · Mistaken Identity · Go A-Maying · Mother's Day · Super Dads · Sea Horse · Midwife Toad · Greater Rhea · Gold-Headed Tamarin · Parent for a Day · Dyeing Eggs—Naturally · You Crack Me Up · Egg Treasures
The End (for This Year)

INDEX   121

IS IT SPRING YET?

IS IT spring yet? Do tree buds look as if they're about to burst? When you go outside, does the sun seem brighter even when the air is frosty cold?

Spring slips up on you just when you've given up on its arrival. It's bright green shoots, peaking out from under lingering lumps of dirty snow. It's icicles dripping, parked cars coated with yellow pollen, a lone butterfly fluttering in search of a flower. Spring smells sweet, drips down your neck, nearly blows you away. It's feeling sweaty in the afternoon after having goose bumps in the morning.

Spring is being surprised. It's eating the first strawberry, spotting the first robin, or discovering the first dandelion in bloom. Sometimes spring is gray, sniffly, soggy. It may make you sneeze, but this season will never bore you.

Is it spring yet?

Almost? Then this book is for you.

It has investigations to help you discover spring and instructions for building things to enjoy during this season. There are stories about how animals win a mate and tales of super animal moms and dads. It has facts about all sorts of babies growing up and hints for starting your own baby-sitting business. There are directions for mixing your own bubble-blowing solution and using natural dyes to color eggs. It has instructions for building your own kite and tips for flying it. You can make a pinwheel, blow on a grass whistle, and hold an egg race. There are games to play, riddles to laugh at, and lots of facts to amaze you.

This book is guaranteed to bring out that secret desire to become an explorer that lurks inside you.

Is it spring yet? Get ready. Spring is too full of surprises to miss.

1. SPRING FEVER

Early Warning Signs

IF YOU live where winter weather is often cold and sometimes miserable, you're probably especially eager for spring. So even though this season doesn't officially arrive until the twentieth of March in the earth's northern hemisphere, nature slips in some signs to let you know it's coming.

Look carefully when the snow first melts, exposing flower beds. The bright green tips of crocuses and daffodils are a clue that spring is approaching. Snowdrops and primroses also appear early. Each year, these plants seem to race to produce springs' first blooms. And check out the buds on willows and birches. These trees are among the first to unfold their leaves each spring.

Then listen on a warmer-than-usual day. You may hear the buzz of insects, the chirp of frogs, or the chatter of squirrels, even if you don't see these animals. Hibernators and cold-weather nappers are waking up.

Watch for robins, too. You can't miss their bright red breasts. If you see just one, that bird may have stayed the winter. An occasional one will. But if you begin to spot robins here and there, you can bet these early birds are back from their winter homes. According to superstition, you can make a wish when the robins come back. Wish for longer days, and you can count on your wish coming true.

Even where the weather is warm year-round, each day now has more hours of sunlight than the day before. You can prove this by keeping a chart of the sunrise and sunset times each day. These are usually printed in the newspaper and announced during radio and television weather reports. If you plot these times on a line graph, you'll be able to see a pattern develop.

You can use this pattern to predict future sunrise and sunset times. Check your predictions by observing when the sun actually rises and sets.

The number of hours that sunlight reaches the earth has a lot to do with making spring a warmer season. Why? Sunlight doesn't warm the air directly as it passes through. The earth's surface absorbs the sunlight and heats up. Then the earth's surface radiates heat, warming the air above it. The more hours the surface absorbs sunlight, the more hours it will radiate heat energy.

The earth's surface not only receives sunlight longer as spring approaches, it also gets a more concentrated energy pack. The earth

is tilted as it orbits the sun. So during part of its orbit, the northern hemisphere is aimed away from the sun, and sunlight strikes the earth's surface at an angle. Then the earth moves into a position where the northern hemisphere is aimed at the sun, and sunlight strikes straight on. Shine a flashlight at a wall, holding it at an angle and then straight on to see the difference. Slanted rays spread out and appear dimmer. Direct rays concentrate bright light on a small area.

RIDDLE

What bird is a thief?
A robin.

For the Record

SPRING ARRIVES every year. But spring weather isn't always the same. Some years this season is warm and rainy. Other years, cold, wintery weather seems to linger until summer. To find out what past springs were like, just check a tree stump.

A tree produces two rings each year—a light-colored spring ring and a dark-colored summer ring. Spring rings are light-colored because trees grow rapidly during this season, producing large cells. During the summer, trees grow more slowly, producing smaller, more tightly packed cells. It's the ring's width, though, that reveals what past weather was like. If the spring ring is wide, the weather was probably warm and wet, encouraging growth. If it's narrow, the weather was probably cold and dry.

Lion Hunting

ONE OF the easiest constellations to spot in the spring sky is Leo. Leo is the Latin word for lion, and that's what ancient stargazers thought they saw in this pattern of stars. To find this heavenly cat, first look for the pointer stars in the Big Dipper. Usually, you look above them to find the North Star. This time look below them. They point to a star triangle in the eastern part of the sky. That's the lion's rear and tail. The stars that form Leo's head and mane look like a big, backward question mark. And one really bright star, Regulus, forms the question mark's bottom dot.

According to an ancient legend, the Greek moon goddess, Selene, became angry with her earth subjects for forgetting their annual sacrifice to her. So she sent a huge, hungry lion to prey upon them. Its skin was so tough no spear, sword, or arrow could pierce it.

At last, the Greek king Eurystheus sent Hercules, the strongest man in the world, to tackle the lion. Hercules fought and struggled with the great beast. And just when it looked like the cat might win, Hercules put a headlock on the lion and choked it to death. Then the moon goddess returned Leo to the heavens, where the cat began its second life.

Uh-oh—look to the northeast above the horizon. Here comes Hercules, chasing Leo. That hunter just won't quit!

Scrub It!

SPRING HAS traditionally been the season to clean up. No one is sure who invented soap or how long people have been using it. But the Romans are thought to have used soap three thousand years ago. And the French are known to have produced soap in A.D. 100. In those ancient times, though, soap was considered a medicine—something to treat sores and skin diseases. The idea of using soap for cleaning didn't become popular until the sixteenth century when people realized that being clean was an important part of staying healthy. Today, soap even comes in different forms for different cleaning jobs—laundry soap, dish-washing soap, bath soap, and more. Each person in the United States alone can be counted on to use about thirty pounds of soap a year.

So how does soap clean? To find out, you first need to discover what happens to water when soap is added to it. Place a sheet of waxed paper on a tabletop. Sprinkle on a few drops of water with an eyedropper or your fingertips. Next, examine the drops closely. You'll see that each drop looks like a little round hill.

Each water molecule is made up of two hydrogen atoms and one oxygen atom. The atoms form bonds, pulling the water molecules together. The molecules in the center of a water drop are being pulled in all directions. This sideways and outward pull resists the downward pull of gravity, preventing the drop from flattening out.

This bonding between molecules also makes the water's surface behave like a stretchy film—a phenomenon called surface tension.

But touch a soapy toothpick to a water drop and watch what happens. Whoosh! The drop immediately flattens out. If you guessed that soap breaks the bonds between water molecules, you're right. Soap molecules have one end that is attracted to water and one end that repels water. This water-repelling end doesn't stay unattached, though. If greasy dirt is present, the water-repelling end of a soap molecule is attracted to it. Scrubbing helps because this motion loosens any dirt clinging to your skin or to the fibers of your clothes. Then more water can be used to rinse the water-soap-dirt molecules away.

Before there were washing machines to agitate the laundry, dirty clothes were soaked in water and then beaten to loosen the dirt. Next, the laundry was dumped into an iron kettle full of soapy water and boiled. This bubbling action also helped loosen the dirt so the soap could attach to it. Finally, the clothes were rinsed several times in clean water.

Does soap and water really clean better than water alone? To find out, cut two one-inch squares of plain cotton cloth. Smear each with one-eighth teaspoon of catsup and let the stain dry. Next, fill two identical glasses half-full of tap water. Add one-fourth teaspoon of a liquid soap to one glass and stir well. Then drop one cloth square into each glass and wait five minutes before removing the squares. Don't squeeze out the cloth. Which looks cleaner—the one soaked in soap and water or the one that was in water alone?

To see what effect agitating has, repeat the investigation you just did. This time, though, add soap to both glasses but stir only one glass for one minute. Then remove both cloth squares and decide which one looks the cleanest.

Clean Fun

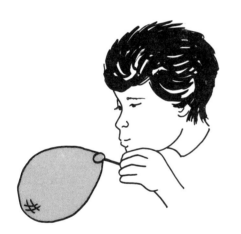

SOAP CAN do more than clean. It can make bubbles. On a spring day when the air is calm, whip up a batch of this Super-Bubble Solution. Then try the bubble challenges. Be sure to notice the beautiful rainbow colors in the bubbles. Sunlight is actually a bundle of different colors of light. As sunlight passes through the bubble, each color of light is refracted (bent) a slightly different amount than the others. This makes the bundle of colors separate and spread out, creating the rainbow hues.

Super-Bubble Solution

Pour one-half cup of liquid dish-washing detergent (one without any additives, such as lemon scent) into one and one-half cups of water. Add one cup of glycerine (available at drug stores). This helps make your bubbles last longer by slowing down the evaporation process, which normally makes bubble film get thinner. When the film is too thin to contain the air pressure inside it, the bubble breaks.

Mix the bubble solution, stirring gently. To blow bubbles, dip a plastic straw into the Super-Bubble Solution, let it drip once, and then blow through the dry end of the straw.

Bubble Challenges

1. What's the biggest bubble you can blow?* (Wonder how to measure a bubble? Let it break against a sheet of paper. Then quickly outline the wet circle and measure the diameter.)
2. What's the smallest bubble you can blow? (Measure the same way you did a big bubble.)
3. Pour three tablespoonsful of Super-Bubble Solution on a sheet of waxed paper and blow into this puddle. Can you blow two identical bubbles that touch? (Notice that the bubbles share one wall where they come in contact with each other. Is this common wall round or flat?)
4. Can you blow three identical touching bubbles?
5. What's the tallest bubble cluster you can blow? (You'll need a friend to help you measure this one.)

*To produce giant bubbles, bend a wire clothes hanger to form a big bubble wand and pour the bubble solution into a 13x9-inch cake pan, or use a large cardboard tube, such as those paper towels come rolled on, as a bubble pipe. Dip one end of the tube into the bubble solution and blow gently into the other end.

Dump It

SPRINGTIME IS also traditionally the season to sort through things and throw out what's no longer wanted. Getting rid of all the trash people dump, though, has become a big problem all year long. You may be surprised to know that the average person in the United States throws away 3.5 pounds of trash every day. That's 1,278 pounds a year per person! And the U.S. currently has about 243 million people.

The way most cities handle all this refuse is to collect it, haul it to an open area, dump it in a pit, and spread a layer of dirt over it. The idea behind getting rid of trash this way is that this material will gradually rot and decompose, returning minerals to the soil. And eventually, when the landfill is full, the area can be used for something else, such as a park. Trash, however, is being dumped so fast that one layer doesn't have time to decompose before several more are piled on top. Also, a lot of what is being dumped doesn't decompose at all.

You can find out what happens to refuse in a landfill by building a model. First, collect an empty, clear plastic two-liter soft-drink bottle and cut the top off. Next, fill this nearly full of soil. Then poke each of these items into the soil next to the plastic wall so you can observe them: a piece of cooked potato, a slice of banana, a piece of paper, a piece of aluminum foil, and a piece of plastic (clear wrap or Styrofoam). Sprinkle the surface of your model landfill with water every other day and write down any changes you observe over the next two weeks.

The Environmental Protection Agency predicts that by 1990, twenty-seven of the fifty states will be completely out of dumping space. New York City has the world's largest city dump. This landfill covers three thousand acres and accepts eleven thousand tons of trash daily. But New York City residents throw away an estimated twenty-four thousand tons of trash daily. So the city must truck away trash to more distant landfills. Philadelphia and Boston have even bigger problems. Those cities have already filled up all the surrounding landfills. So they must haul their refuse great distances—sometimes by barge to places as far away as Panama.

RIDDLE

Why did the dog lie out in the sun?
He wanted to become a hot dog.

Reuse It

ONE WAY to stop this growing mountain of refuse is to find ways to recycle. Some states already require reusable household trash to be separated into containers reserved for plastic, metal, glass, and paper. Then this trash is hauled to special plants where it's heated or treated in some way to regain the basic material. You can try some recycling right at home in your kitchen, though, to make new paper from old. Just follow these steps:

1. First, shred 60 sheets of newspaper as fine as possible, put the scraps in the sink, and cover with water. Let the paper soak overnight to wash out the inks.
2. Scoop the washed paper into a colander a little at a time to drain it. Put 8 cups of this mush into a large tub or dishpan and add 3 cups of water. Stir with a large spoon until the mixture is smooth. Then add one gallon of water and stir again.

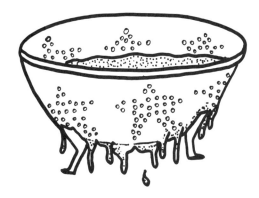

3. Slide an 8-inch square piece of window screen, edge first, into the liquid until it's completely submerged. Lift straight up, letting the pulp mixture drain for about thirty seconds.
4. Lay the screen down on a counter, place several sheets of dry newspaper over the pulp, and flip the whole set up over on the counter. Sponge off any excess water and lift off the screen.
5. Place another couple of sheets of newspaper over the pulp and roll with a rolling pin. Then let the paper set overnight or until all the layers are completely dry. Once they're dry, the sheets of newspaper will easily lift off, and you'll be able to peel off the recycled sheet of paper.

Spring Tonic

REFRESHING THE house wasn't enough. People used to believe that the body needed something extra for spring, too. Their solution was to give everyone a dose of tonic. Usually this was a cup of strong tea made by boiling dried sassafras root bark in water and then sweetening it with honey. If you can find sassafras bark in a store, you'll need to put one cup of shredded bark in one quart of water. Boil for ten minutes and then strain off the tea. Or you could simply add one tea bag of your favorite tea to one cup of boiling water and let it steep (sit covered) for one minute. Then sweeten this with honey. Either way, March is the month to drink it. According to an old saying: "Drink sassafras tea during the month of March, and you won't need a doctor all year."

Berry, Berry Good

AFTER A winter of dried and preserved foods, juicy strawberries were a sweet treat people eagerly hunted for in the spring. Originally, they were called "strewberries" because the clusters of bright-red fruit seemed to be scattered among the leaves of the low-growing plants. And while it was their taste that made them popular, strawberries were also a good source of vitamin C and iron. It took finding a whole bed of these plants to get more than a taste, though, because wild strawberries are small. Today's varieties are giants by comparison. And commercially-grown hybrid plants supply grocery stores with strawberries year-round.

You won't have to shop for your spring treat, though, if you grow your own. And you can do that even if you live in the city. Start by purchasing what is called a strawberry planter jar at a store that sells garden supplies. This jar looks like a clay pot with cuplike pockets around the sides. Fill it by covering the bottom with a layer of pebbles or small rocks. Next, pour in enough potting soil to fill the jar half-full, pat down the soil, and then finish filling it to the top. After you've patted down the soil again, pour in a half-cup of water and let the jar set overnight. In the morning, add more soil if necessary so the pot and the pockets are full. Then plant a strawberry plant—available from garden-supply stores or plant nurseries—in each pocket. Be sure each plant is buried deeply enough that its crown of leaves rests on the surface.

Place your strawberry pot in a sunny location and add water to the top of the pot as needed to keep the soil moist but not soggy. If you plant early in the spring, you should see blooms in May. But you'll need to pay attention to the weather report. If there's any chance of frost, cover your strawberry pot with a plastic garbage bag.

You'll probably only get enough strawberries to enjoy eating them freshly washed. For a special treat, though, you may want to melt a cup of semisweet chocolate chips in a double boiler and dip the tip of each berry. Set these—undipped end down—on waxed paper until the chocolate has cooled and hardened. Then eat.

Big Hairy Deal

ANIMALS THAT live where it's cold in the winter shed their heavy coats in the spring. A good example is the musk-ox. With two-foot-long guard hairs forming a shaggy outer coat and dense underwool acting like insulated long johns, these animals are well-dressed for roaming the super-cold, windswept Arctic tundra. As the weather warms up in late May and June, the musk-ox begins to shed its soft underwool.

You probably know that farmers shear sheep in the spring to collect their wool. But you may be surprised to know that herds of musk-oxen are also kept for their wool. Called *qiviut*, musk-ox wool is softer than cashmere and eight times warmer than the same weight of sheep's wool. So in Alaska, a group called the Oomingmak (native word for musk-ox) Musk-Ox Producers Cooperative keeps a herd of these hairy beasts. And when they start to shed, handlers use long-toothed combs to collect the underwool. It takes about two to three hours to completely comb out one animal. One adult musk-ox yields about six pounds of qiviut. That may not sound like a lot, but only four ounces are needed to knit an average-sized dress. And one pound is enough to spin a thread ten miles long.

Let's Molt Again

BIRD FEATHERS may seem tough, but flying is hard work and causes feathers to break. So at least once a year, birds lose their feathers and grow new ones. This process is usually gradual, though, so birds aren't left naked waiting for their new coat. And the feathers are normally shed in a set pattern from tail to head with flight feathers being lost in pairs—

one from the right wing and the corresponding feather from the left wing. Ducks and geese are an exception, shedding most of their flight feathers all at once. So these birds are briefly grounded.

Many birds shed twice a year—once at the end of the summer nesting season and once in the spring. Females keep a generally drab appearance year-round so they can nest without being noticed by predators. But the males often develop "fancy" feathers in the spring just in time to attract a mate. These may just be brighter versions of its normal colors or the change may be more dramatic. The male weaverbirds, for example, have brown feathers most of the year but sport a coat of blood red feathers for courting. The male ruff develops a huge, feathery frill around its neck. And while the male puffin's feathers stay unchanged, it develops a brightly colored crest on its beak, and its feet change from dull brownish gray to bright red. It's no wonder they like to show off after they've molted!

Let's Play

WANT SOMETHING fun to do outdoors when the weather is wonderful? Here are two games that are old-time favorites.

Hopscotch: Use chalk and a measuring tape to draw the court shown on the next page on smooth, level cement or asphalt. Any number can play. Each player will need a flat stone—no bigger than 3.5 inches in any direction—to use as a "puck." Begin each turn behind the baseline with the puck in one hand. Then follow the rules for each stunt— one at a time—trying to complete one stunt per turn. Your turn ends any time the puck doesn't land in the designated section of the court or anytime the puck or your foot touches one of the court lines. If you don't successfully complete a stunt during your turn, you must try that stunt again during your next turn. The stunts are complicated, but the challenge is what's fun. Just read the directions for one stunt at a time or have someone read them to you as you perform the activities. The winner is the first person to successfully complete the most stunts after everyone has had an equal number of turns. Good luck!

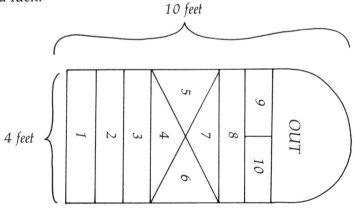

STUNTS

1: Toss or drop the puck into square 1; hop into square 1 on one foot; kick the puck out across the baseline; and then hop over the baseline.
2: Toss or drop the puck into square 2; hop into square 1 and then into 2; kick the puck out across the baseline; hop back into square 1; then hop out across the baseline.
3: Toss the puck into section 3; leap, landing with the right foot in square 1 and the left foot in square 2 at exactly the same time; jump on one foot into section 3; kick the puck out across the baseline; if it stops, leap back, straddling 1 and 2 with your right foot in 2; raise either foot, and while hopping, kick the puck out; then hop into 1 and out across the baseline.
4: Toss the puck into section 4; follow the same procedure you did to get to 3 only hop into 4; then kick the puck out and exit across the baseline as before.
5: Toss the puck into section 5; follow the same procedure you did to get to 3 in order to move to section 5; retrieve the puck; then hop out across the baseline.
6: Toss the puck into section 6; leap into 1 and 2, landing with your right foot in 1 and your left foot in 2; jump to 3, landing on one foot; leap, landing with your right foot in 4 and your left in 5; jump to 6, landing on one foot; retrieve the puck as before by hopping and kicking it out. To exit, leap, landing with your right foot in 5 and your left in 4; jump to 3, landing on one foot; leap landing with your right foot in 2 and your left in 1; then jump out and land beyond the baseline, standing on one foot.
7: Toss the puck into section 7; advance as you did in stunt 6 but land on both feet in 7; kick out the puck and exit as you did in stunt 6.

8: Toss the puck into section 8; advance as you did in stunt 7; hop into 8 on one foot—either foot; kick out the puck and exit by landing on both feet in 7, then proceeding as you did in stunt 6.

9: Toss the puck into section 9; advance as you did in eight; stay in 8, leaning over to pick the puck up from section 9; exit as you did in stunt 8, carrying the puck in your hand.

10: Toss the puck into section 10; advance as you did in 9, hop into 10 on one foot; pick up the puck; hop into 8, and exit as you did in stunt 8, carrying the puck in your hand.

11: Don't use the puck on this final stunt. Advance to 8 as you did before; leap to land with your right foot on 9 and your left on 10; do an about face by doing a leaping half-turn; jump, landing on one foot in 8; then exit as before.

RIDDLE

What bird is there every time you eat?

A swallow.

Marbles: Start by drawing a chalk circle about eight feet in diameter on smooth, level cement or asphalt. Or scratch a circle in bare dirt with a stick. Next, draw two parallel lines—each touching opposite sides of the circle.

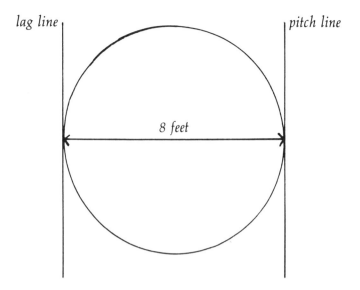

Any number can play. To determine playing order, each person must pitch or shoot a marble from the pitch line (line closest to you) toward the lag line (line on the opposite side of the circle). The person whose marble comes the closest to the lag line goes first, the next closest goes second, and so forth.

To shoot, rest your hand on the ground, place the marble against your index finger, and propel it away with a flick of your thumb. The marble used to shoot is called the shooter, the taw, or the moonie.

Once the shooting order is established, each player places the same number of marbles inside the circle. These may be arranged in a bunch in the center. Or a large cross may be drawn in the center of the circle, and the marbles may be arranged along this design. Each player begins his or her turn by shooting at the target marbles inside the circle from any point on or outside the circle. The goal is to knock the target marbles outside the circle. If a marble is knocked out, the player collects it, scores one point, and continues, shooting from the point where his or her taw stops. A miss ends the turn. If a marble is only knocked partway out, it's counted as a miss but the marble is left where it is.

When all the marbles have been knocked out of the ring, the winner is the one with the most points. Then the marbles that have been collected are sorted and returned to their owners.

Say, is anyone in your neighborhood building a house? You may be surprised to discover just how much construction is going on. Spring is the season when many animals set up housekeeping, too. You'll find out about these families and discover how to be a good neighbor in the next chapter. Read on.

2.
FAMILY TIME

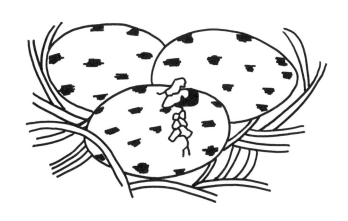

FOREST GLEN ELEMENTARY SCHOOL
1935 CARDINAL LANE
GREEN BAY, WI. 54313

Going Courting

WHEN ANIMALS instinctively sense that it's time to start a family, males often go to great lengths to attract and win a mate. Some offer special gifts. Others do their best to look bigger and better than anyone else. They may even fight battles to prove themselves. And still others chase after the object of their affection. Where the weather is generally warm all year, this family feeling may come at any time. But for animals living where there are definite seasons, families are usually started in the spring. Then the young are born when the weather is less harsh and plenty of food is available.

Bowerbirds

MALE BOWERBIRDS must believe it takes more than fancy feathers to attract a mate. These residents of New Guinea and Australia try to outdo each other by building elaborate courting spots called bowers. Construction begins when the male clears leaves and twigs from a circle three to five feet in diameter on the forest floor. Next, this site is decorated with brightly colored berries, shells, or flowers.

Some species of bowerbirds build around a small tree. Then they add a hut, piling branches against the tree trunk until the structure is about twenty inches high. Withered flowers and rotting berries are regularly plucked from the tidy bower and replaced with fresh ones. All day long, the male sings and poses proudly in his bower. If a female joins him, they mate. Then the female goes off alone to build a nest high in the treetops.

Snowshoe Rabbits

THESE MALES have to be fast to catch a mate. During March, their mating season, a female expresses her interest in an approaching male by suddenly leaping into the air and darting away. Then the pair races uphill and down at amazing speeds of up to thirty miles per hour.

Hummingbirds

IT MUST take sensational aerial displays to win a female hummingbird. Why else would a male hummingbird swoop in circles and zip back and forth in front of the female of his choice? Then, for a grand finale, he soars nearly sixty feet into the air, turns, and dives straight back down. Just before crashing into her, the male brakes and hovers before his dazzled lady fair.

male hummingbird

Pigeons

THESE MALES have learned that the best approach is to play hard to get. First the male finds a nesting site. Then he settles down and begins to call: "Roo roo roo." Periodically, he will also puff up his neck, spread his tail, and turn around. Eventually, all these actions attract a female. But when she approaches, the male charges, turning her away. This seems to make the female even more interested because she returns. If the female persists, the male pigeon will finally accept her. And to show his approval, he marches behind her, herding her along. If you live near a park, you may want to visit to observe the pigeons courting in the spring.

Grizzly Bears

THE TOP bear gets the most females. So adult male grizzlies challenge each other and fight to establish rank. These battles go on year-round but are most common in the spring, during the mating season. And since male grizzlies may be seven or eight feet long and weigh as much as eight hundred fifty pounds, the battles often leave both the loser and the winner badly scarred. The big bears lunge at each other, slapping with their powerful forelegs and long claws. Then they lock jaws and wrestle until one of the opponents gives in.

Although grizzlies mate in the spring, the baby bears don't actually begin to develop inside the female until she hibernates the following fall. And they're born—blind, helpless, hairless, and weighing only eighteen ounces—just six to eight weeks later. They nurse snug in their mother's den all winter. So when the mother and cubs emerge in March or April, the young bears are six-pound balls of energy.

Siamese Fighting Fish

IF YOU have a tropical aquarium, you might want to watch a pair of these fish. When ready to mate, the normally brownish gray male develops dazzling, iridescent colors. Then he builds a floating nest of bubbles on the surface, takes up a defensive position under this nest, and rushes out to meet any intruder. If two male Siamese fighting fish meet, they ram each other—jaws wide open—repeatedly, until one retreats. If the intruder is a female Siamese fighting fish, she is swept off to the nest. The eggs that are laid and fertilized there begin to sink, but the male quickly gathers them in his mouth and blows them back into the bubble nest.

I Love How You Smell!

PEOPLE AREN'T the only ones who use perfume to attract the opposite sex. Lots of insects use scent signals. Scientists call these *pheromones*. Male moths, for example, are able to smell with their feathery antennae. And they can locate the scent of a female moth of the same species even if she's several miles away. Because many

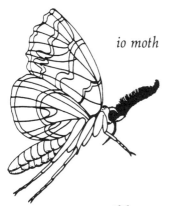
io moth

insect pests use scents to find a mate, the U.S. Department of Agriculture has begun to make this a fatal attraction. Traps are baited with pheromones to catch insects before they can reproduce.

Falling in Love

EAGLES CHOSE one mate for life. But each year when they return to their nesting site, the eagle pair renews their family ties by repeating their dramatic courtship ritual. First the male and female fly toward each other again and again—barely missing collision—as they climb upward. They dip and turn, soaring in large circles high in the sky. Then, suddenly, the pair comes together, locking their feet. And with their wings partially tucked, the eagles plunge toward the earth, tumbling over and over. Just in time, the male and female eagles separate, swooping upward again.

No Trespassing!

SOMETIMES PEOPLE put up fences and post signs to warn intruders off private property. Animal boundary lines may not be so easy to spot, but to the animal defending its territory, the area is just as clearly marked. Why do animals fight over land? It's because there is only so much food available, and animals must compete for this food in order to survive.

Some animals that live in groups, such as wolves, stake out and defend a territory all the time. Other animals only claim a territory when they're preparing to have a family and therefore need a dependable food supply close by. The boundaries of the group's or individual's territory may be marked with strong smells. That's why dogs spray their urine on a variety of objects. Animals, such as certain antelope, have special scent glands they rub against objects to mark their territory. Bison are an example of animals that use visual markers. These big animals scar tree trunks with their horns to establish their territory.

Songbirds announce ownership with a song. If you have robins in your area, watch them in the spring. A male bird, which can be identified by its brighter breast feathers, will stake out a claim as big as half an acre. Then the proud landowner will patrol the perimeter of his territory, repeatedly stopping at selected perches to sing. Try to observe one male on his regular rounds. Can you map out that robin's territory?

Animal Nurseries

FOR SOME animals this is a safe place to leave their eggs to develop and hatch on their own. For others it's a home where the babies can

be born and grow up in safety. For still others it's a temporary resting place while the young are growing. There are even some parents who come equipped with a built-in nursery. Whether temporary or long-term, nurseries are where baby animals get their start.

American Alligator

BIRDS AREN'T the only nest builders. An alligator mother builds a nest for her young, too. First, she uses her jaws to tear up grass, twigs, and leaves. Then she piles this material together—packing it down with the weight of her body—until it forms a mound two feet high and five feet wide. Then mom scoops a hole in the middle of this nest, lays as many as sixty eggs, and seals over the opening. Many reptiles abandon their young once the eggs are laid, but an alligator mother stays close by, guarding the nest from egg snatchers. Meanwhile, the decaying plant matter in the nest gives off enough heat to keep the eggs at 83°F—the temperature needed for the babies to develop. After nine weeks, the baby alligators hatch. Hearing their soft grunts, the mother gator tears open the nest. The emerging babies quickly rush toward the water.

Skunks

EARLY IN the spring, the mother skunk digs a tunnel about two feet into the ground. At the end of this, she hollows out a burrow and lines it with dry grass and leaves to turn it into a snug nursery. Skunk babies are tiny, furless, and helpless at first. The babies grow quickly, though, and by the time they're seven weeks old, they're strong, fur-covered bundles of energy. Then mother takes them with her at night when she goes out hunting for food. The babies follow their mother single file, staying close enough for her to protect them from danger.

Opossums

THE FEMALE opossum may take shelter in a deserted den or a hollow tree when her babies are born, but the opossum mother has a pouch that is the real nursery. At birth, baby opossums are furless and no bigger than a kidney bean. They crawl into her fur-lined pouch and attach themselves to a nipple. Then they hang on while the mother pumps milk into them regularly for the next couple of months. Even after they're about the size of mice and strong enough to leave the pouch, the baby opossums cling to their mother. They ride on her back, entwining their tails with hers. And at night they crawl back into her pouch to nurse and sleep.

Deer

DEER HAVE lots of enemies, so a mother deer only stays with her baby long enough to nurse during the first two weeks. As soon as the fawn is fed, she moves away. The baby deer—camouflage-spotted and lacking the characteristic deer scent—is safer alone. Later, when the fawn is stronger, it will follow its mother, searching for food as she grazes during the night. Then if danger threatens, the doe will lead the fawn away, racing into the safety of the dense woods.

fawn

RIDDLE

What insect would make the best outfielder?

A spider—it's great at catching flies.

Become a Bird Landlord

PURPLE MARTINS make good neighbors because they eat a lot of flying insects. These birds like to live together in colonies, but unlike many other birds, they aren't messy. They carry their droppings away from their home area in little capsules. So if you'd like to have purple martins in your backyard, you can put up an apartment house. One of the simplest purple martin houses to build is made out of gourds. You'll need to buy at least ten fat, round gourds with short necks. If you're willing to make this a two-year project, you may want to buy seeds and grow your own gourds.

Once you have the gourds, drill a round door hole two inches in diameter in the side of each gourd. (You may want an adult's help.) Shake out the seeds. Next, drill two small holes on opposite sides of the neck, cut a piece of nylon cord ten inches long, thread it through the holes, and tie the cord to form a loop. You'll use this to hang up the gourd. Finally, drill two small drainage holes in the bottom of the gourd. You can hang these gourds about twenty feet above the ground in the branches of a dead tree. Or you can put up a pole that's twenty feet high, nail two boards across the top, and hang the gourds from these arms. (Be sure to ask for an adult's help with this part of the project.)

Purple martins prefer an open area close to a pond or a lake. The reason they like to be close to a body of water is that this provides a breeding area for many of the insects they eat. The nesting gourds need to be put out in February. Early in March, purple martin scouts will arrive from the birds' winter home in South America, looking for nesting sites. You can expect pairs to move into your apartment house late in March.

Build a Robin's Nest?

YOU CAN build a nest the way a robin would to see how this bird constructs its treetop home. First, cut a piece of heavy cardboard to work on. Draw a circle six inches in diameter on the cardboard. That's how big a robin's nest usually is. Collect enough blades of grass and pine needles to cover this circle. Only carry a couple of blades of grass or pine needles at a time to your nesting site, though. That's all a robin can carry. This will help you appreciate the amount of effort the bird must make.

Next, carry mud to your nest—one-fourth teaspoon at a time. Press these scoops of mud into the plant material, working it in with your fingertips, to create a kind of cement. If a robin can't find mud, it will scoop up a beakful of soil and then stop to dip this in water, making mud. So if mud isn't available, you can make mud this same way, mixing soil with water until it's the consistency of thick pancake batter.

Once you have the nest base finished, build the walls up slowly all the way around about a half inch or less at a time. Use pine needles toward the inside and small twigs toward the outside, cementing this material together with mud as you did the base. The mud should be such an integral part of the nest that you don't notice it from the outside. As the walls grow higher, smooth the inside with you hand to create a cup shape. A robin does this by sitting on the nest while turning and twisting. The finished nest walls should be about six inches high. Finally, line the nest with soft grass to make a pad for the eggs.

RIDDLE

Why did the bee call the flowers lazy?
Because they were always in their bed.

Home Sweet Nest

IF YOU thought building a robin's nest was hard, you'll really be able to appreciate these bird homes.

You could sit in a bald eagle's nest. An eagle pair add on to and repair the same nest year after year. Eventually, the nest may be nine feet across, fill the crotch of a tree twenty feet deep, and weigh a ton.

A hummingbird's nest is no bigger than a thimble. This tiny bird uses lichens and moss. Then it cements everything together with spider webs.

You'd think building a nest would give a pileated woodpecker a headache! To get started, this bird must first use its beak to chip a three-foot tunnel into a tree trunk.

Ever tried to tie a knot using only your teeth? That's not much different than what the Baltimore oriole does as it builds its nest. First, long strands of plant fiber, hair, or string are draped between two branches and anchored with strong knots. Then more strands are used to weave a pouch. The finished nest is a cradle hanging in a tree top.

Help a Bird Build a Nest

BIRDS NEED lots of building material and you can help supply some. Collect the hair that accumulates in your hairbrush. Then hang this outdoors on a twig. Birds like scraps of string and yarn, too. Don't make the pieces too long, though. Remember the nest builder will need to fly home with its construction material in its beak.

House for Dinner

THINK YOU'VE got an appetite big enough to eat a house? That's what you'd be eating if you had birds' nest soup. The Chinese think this soup is a special treat. They use the nests of one variety of swift to prepare it. These swifts construct their nests entirely from their own saliva. The nests are pearly white and become as soft as jelly when soaked in water.

All of this wooing and nest-building effort has been made to get ready to produce new life. Some youngsters will be on their own as soon as they're born. Others will get years of parental attention. It's part of nature's plan that these babies are too cute to resist. You won't want to miss the next chapter. It's all about animal babies.

3.
HAPPY BIRTHDAY, BABY

Having Babies

LIFE BEGINS for the embryo (developing young) at the moment a sperm cell, the male reproductive cell, joins an egg or ovum, the female reproductive cell. One and one normally equals two, but in fertilization—the fusing of a male and a female cell—the results are one—one new life.

The ovum and the sperm cell each contain chromosomes, chemicals in the cell that tell the cell what traits the embryo will develop. Since these chemicals from the mother's egg cell and the father's sperm cell are combined, the embryo will be somewhat like its mother and somewhat like its father.

Fertilized eggs may be jelly-coated like fish eggs, frog eggs, and salamander eggs. They may have leathery cases like turtle eggs and snake eggs. Or they may be hard-shelled like birds eggs. But all eggs serve the same purpose. They are life packages, supplying food, oxygen, waste disposal, and protection during the baby's early stages of development.

Can you identify what kind of animal will grow from each of these eggs? Check yourself by reading the answers upside-down below the pictures.

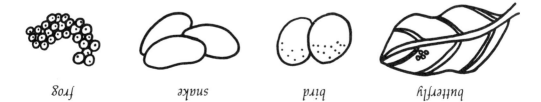

frog   snake   bird   butterfly

Laying an Egg

IN THIS picture you can see how an egg is produced inside a bird's body.

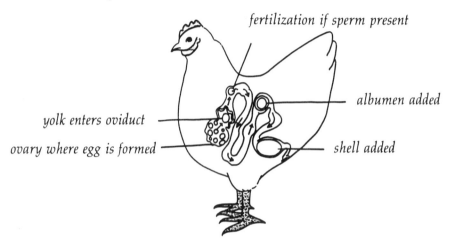

Eggsamination

EXAMINE A raw chicken egg. The first thing you'll see is the hard shell. When the egg was first laid, this shell was covered with a protective layer called the bloom or cuticle. This bloom was a mucuslike layer that filled all the tiny pores in the egg shell to keep bacteria and mold from entering the egg.

Are you wondering how anything could get through an eggshell? It's not as solid as it looks. Submerge the egg in a glass of warm water. You'll see lots of tiny bubbles appear on the outside of the shell. Air is escaping through the thousands of small openings, or pores, in the egg's shell.

To see what's inside the egg, crack it and pour the contents into a clear plastic cup. First, find a tiny white patch on the yolk. This jellylike patch contains the ovum. The eggs you buy from the grocery store usually aren't fertilized, but if this ovum had been fertilized by merging with a sperm, cell division would have begun immediately, creating an embryo.

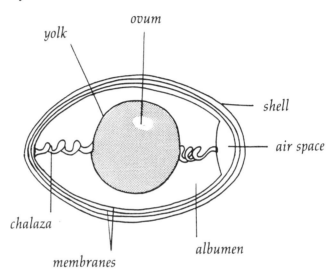

The thick, white, ropelike strand on either end of the yolk is called a chalaza. They hold the yolk steady. If the egg rolls, the yolk will roll, too, keeping the embryo on top—the best position for the baby bird's development.

The yolk supplies the food for the chick. So if an embryo had been developing, the yolk would be covered with a network of blood vessels, connecting it to the embryo's system of blood vessels. Gently touch the yolk to see that it's covered by a thin membrane.

You'll find two more membranes lining the egg shell. The thicker membrane can be peeled away easily. You can only see the thinner membrane if you tease bits of the shell apart. These membranes allow the oxygen needed by the embryo to enter the shell and the

waste carbon dioxide gas given off by the developing chick to escape.

If an embryo were growing in the egg, two more membranes would be present. A fluid-filled membrane, the amnion, would envelope the embryo and protect it from injury due to sudden blows and temperature changes. The other, the allantois, would form a thick network of blood vessels close to the shell to handle the embryo's exchange of oxygen and carbon dioxide with the outside world. The allantois would also store waste products produced as the embryo uses the food from the yolk.

The albumen is the clear, jellylike material surrounding the yolk. It's usually called the egg white because it turns solid and white when cooked. The albumen provides an additional supply of food for the embryo.

A chick requires twenty-one days to develop. After eighteen days of development, most of the yolk will be used up. The chick's body will absorb any leftover yolk before it hatches. And the baby bird will swallow any remaining albumen.

Finally, look at the empty shell. In the rounded end, you'll find an air pocket about the size of a dime. When an egg is first laid, the contents are very warm and completely fill the inside of the shell. But as the contents cool slightly, they shrink, and air is drawn in through the tiny holes in the shell to fill this space. If the egg is fertilized, the embryo develops with its head near this air pocket. And just before it breaks out of its shell, the chick pokes its beak into this pocket, drawing its first breath.

RIDDLE

Why didn't the little boy use toothpaste?

His teeth weren't loose.

A Naked Egg

THE SHELL of a chicken egg is 93 to 98 percent calcium carbonate. You can remove this hard shell without cracking the egg. Simply submerge it in a glass full of vinegar and let it sit overnight. Vinegar is a weak acid and will completely dissolve the shell. Slide the naked egg out onto a plate. Then touch it and take a close look at the egg enclosed in its membranes.

Eggsperiments

YOU CAN find out some things about an egg without cracking the shell. For example, want to know how fresh it is? Submerge the egg in a quart jar two-thirds full of water. You learned that the air pocket inside the egg forms as the contents shrink after the egg is laid. As the egg gets older, some of the liquid inside begins to dry out. Then even more air is drawn in, increasing the size of the air pocket. That's why a fresh egg will lie flat on the bottom of the jar; a slightly stale egg will have one end tipped up—the one with the air sack; and a very stale egg will be suspended almost vertically.

Or want to find out if an egg is raw or hard-boiled? Try to spin an egg on its end on a flat surface. The liquid center of a raw egg creates a drag that makes spinning impossible. Only a hard-boiled egg will spin on its end.

Good Egg

EGGS ARE designed to provide a developing embryo with all the food it needs. Eggs are also a good food source for people, although they don't supply everything a human needs to live and grow. Eggs lack vitamin C, calcium (since people can't easily digest the shell), and carbohydrates. They are, however, an excellent source of protein.

Eggs also make an easy-to-carry snack food once they're hard-boiled. They even come prewrapped in a shell. To hard-boil a batch of eggs, place four to six eggs in a one-quart saucepan. Cover with cool water. Then put the pan on the stove and switch the burner on to medium-high. As soon as the water comes to a rolling boil, take the pan off the burner and cover it with a lid. Let the eggs sit—covered—in the hot water for twenty minutes. Then use a spoon to lift the eggs out of the water, set them on an oven rack, and let them cool.

Off to the Egg Races

ARE YOU ready for a challenge? Then use string to mark off a lane eighteen inches wide on a flat grassed area or use chalk on level pavement. Also mark a starting line at the beginning of the lane and, twenty feet away from this, a finish line. Next, place a hard-boiled egg on the starting line,  get down on all fours, and push the egg toward the finish line with your nose. Time yourself to see how fast you can complete the course.

This race is harder than it looks. Although chickens are now usually raised in special environments so their eggs can be collected, they're related to ground-nesting birds. And like other ground-nesting birds, their eggs are oval-shaped—a shape designed to make an egg wobble rather than roll away from a nesting site. Tree-nesting birds generally lay round eggs.

When you become an expert egg roller, challenge your friends to a race. The most famous egg-rolling event in the United States is traditionally held on the White House's south lawn on the Monday following Easter. In this event, though, the eggs are pushed with spoons. The custom of rolling eggs supposedly started on the Capitol grounds about 1810 and was moved to the White House lawn by President Rutherford B. Hayes in 1878.

Rare Eggs

ONE OF the most expensive foods in the world is fish eggs or roe. To gourmets, the eggs of sturgeon and other large fish is royal de luxe caviar. And the going price is about $500 for 500 grams—a little less than a pound.

RIDDLE

What happened when the nail had a fight with the tire?

The nail knocked it flat.

Tough Stuff

YOU PROBABLY won't be surprised to learn that ostriches, the world's biggest birds, produce the world's biggest bird eggs—about the size of a softball. But you may be amazed by how strong an ostrich egg is. The shell is so strong that an adult man can stand on the egg without cracking it. This also means, though, that an adult ostrich, which can weigh as much as 340 pounds, is able to incubate the egg without squashing it. If you think you'd like to eat a hard-boiled ostrich egg, you'll have to be patient. It takes about forty minutes for one egg to cook completely.

Mermaid's Purse

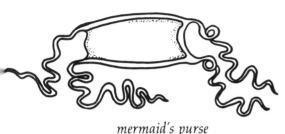

mermaid's purse

THESE YELLOW, red, black, or brown pouches feel like they're made of leather. But if you ever find one washed up on a beach, peek inside. You'll be in for a big surprise. This purse-shaped pouch will contain a baby shark.

The long, threadlike tendrils that resemble purse strings at the corners wrap around rocks or seaweed, anchoring the egg while the baby shark develops. Depending on the kind of shark, this may take from six to fifteen months.

Whale sharks, who as adults may be as long as a bowling alley, produce the largest mermaid's purses. These egg cases are about twenty-seven inches long and sixteen inches wide. Use a yardstick and tape or string to outline an area on the floor as big as one whale shark's egg. Then sit in the space. Would you have lots of room inside this mermaid's purse?

Giving Birth

IN SOME animals, the embryos grow and develop inside the mother's body. When these babies are born, they are small and often need a lot of care. In the picture on page 54, you can see baby kittens growing inside their mother.

A special tube, the umbilical cord, carries blood with oxygen and food from the mother's body to each baby and carries away carbon dioxide gas and wastes.

> RIDDLE
>
> What do polite lambs say to their mother?
>
> *Thank ewe.*

Mother's Milk

SOME ANIMAL mothers have special glands that produce milk to feed their babies. These animals are called *mammals*. People are mammals, too. So your first food may have been your mother's milk.

Although the milk each different type of mammal produces is slightly different, it always contains these ingredients: water, sugar, fat, protein, minerals, vitamins, and antibodies.

The sugar in milk is called lactose. Babies need lactose for energy, to stay warm, and to grow. Kangaroo milk has the most lactose. So if you drank kangaroo milk it would taste very sweet. Cow's milk has slightly less lactose than human milk. And seal's milk has the least.

Milk appears white because the protein, called *casein*, and the fat don't dissolve in the water the way the other ingredients do. Instead little tiny droplets of the casein and fat remain suspended in the water. The protein in milk helps babies grow muscles and other body tissues. The fat is stored as extra food for energy. Mothers, such as whales, that live in very cold places produce milk rich with fat to help their babies stay warm.

The most important mineral in milk is calcium. Babies need this to produce strong bones and teeth. Milk contains many vitamins, but vitamin D, like the calcium, is particularly important for growth.

Antibodies help the body fight off diseases. Some animals, such as horses and pigs, are born without antibodies. They must get all their antibodies from their mother's milk until their body begins to defend itself. Human babies do have some antibodies when they're born, but the antibodies in their mother's milk add even more protection.

Fast Baby Food

WHALE BABIES must nurse underwater, and they can't hold their breath as long as an adult whale. So as soon as the calf attaches to one of the two mammary glands close to its mother's tail, milk begins to be pumped into the infant. The baby receives nearly three gallons of milk in less than five minutes. A whale calf nurses about forty times a day, so it's no wonder it weighs over three thousand pounds when it's only six months old!

An Orange Moo

MILK IS such a nutritious food that a lot of people drink it as adults, too. In fact, the average American drinks almost one-hundred quarts of milk a year.

If you'd like to try a new, tasty way to enjoy milk, collect a tall glass, a long-handled iced-tea spoon, an ice-cream scoop, vanilla ice cream, frozen orange juice concentrate, milk, and a straw. Then follow these directions:

1. Spoon two teaspoons of the frozen orange juice concentrate into the glass.
2. Add two scoops of vanilla ice cream.
3. Fill the glass with milk to within two inches of the top. Then stir slowly to mix the orange juice concentrate with the milk and ice cream.
4. Put the straw in the glass and drink up.

What's the Baby's Name?

YOU KNOW baby cats are called kittens and baby dogs are called puppies. So what are the names for these animal babies? Check yourself by reading the answers upside down below the pictures.

swan/cygnet shark/cub kangaroo/joey frog/tadpole or polliwog skunk/kitten deer/fawn

Growing Down

BABIES ALWAYS start out little and grow bigger as they become adults—right? That's what you'd expect, and it's usually true. The South American paradoxical frog does just the opposite, though. The tadpoles of this kind of frog are about thirteen inches long. But as the tadpoles change into adults, they shrink. Adult paradoxical frogs are only one and one-half inches long.

Nose Job

AT FIRST, a baby elephant just finds its trunk a big nuisance to be curled up out of the way while it feeds on its mother's milk. Mother elephant caresses her baby with her trunk and twists her trunk around her little one in an elephant kiss. When the youngster misbehaves, she gives it a nosy spank.

Soon the elephant baby discovers, though, that a trunk can be very useful. It provides a movable sense of smell. Without turning its head, the youngster can sniff in any direction. And it makes an excellent snorkel, letting the baby breathe while swimming underwater. The fingerlike tips of its trunk can also be used to pick up leaves and fruit. When the baby elephant remembers to suck in, a trunk can even be used to draw in water, fine sand, or soil with the force of a vacuum cleaner. Then the trunk becomes a hose to blow out a spray of cooling water or soothing dust on the baby's sensitive skin.

Taking Care of Baby

LIFE IN a wolf pack is hard. So only the dominant male and female, who are the pack leaders and strongest members of the pack, mate and produce young. They may have as many as six puppies. But the care of the wolf pups becomes the cooperative effort of every member of the pack. In fact, after about the first eight weeks of their life, the mother rejoins the hunt, leaving her pups to be cared for by another member of the pack. Males as well as females share this duty. The wolf pack hunts over a territory that may cover many square miles. But while the pups play and then sleep, their at-home sitter stays alert to signs of approaching danger.

Sometimes, mom also will need a break from the pups' rough and tumble play at the den site. Another member of the pack will babysit then, too.

When a baby dolphin is born, it must quickly reach the surface to take its first breath. Other female dolphins gather close to the mother as the baby is born. Then they swim under the baby dolphin, carrying it to the surface if it needs help. As the dolphin grows, all the adults in the school act as sitters from time to time, protecting and playing with this water baby.

Baby elephants also have sitters, and it's a good thing. While many animals have a very short childhood, elephant calves don't reach maturity until they are between twelve and sixteen years old. In fact, even though it starts to graze when it's only a few months old, an elephant baby continues to feed on its mother's milk until it's about four years old. With this big baby needing so much attention, it's lucky that there are always one or more females in the elephant herd to act like devoted aunts. These females watch over the elephant calf while the mother grazes or bathes. Other mother elephants in the herd will also baby-sit—even letting their temporary charges nurse when they're hungry.

The Baby-sitting Business

THINK YOU might like to baby-sit as a way of making some money? Then here are some suggestions that can help you handle the responsibilities and be successful.

1. Baby-sitting is a real job, so handle it like a business. Keep a calendar as a reminder of when you're due to baby-sit, jotting down the employer's address, phone number, when you should arrive, and when your parents can expect you home. This will help you avoid conflicts and let you be sure you arrive on time. You may also want to use this calendar to keep track of how much you earn.
2. Be organized. Start a card file to keep track of information about each family you work for. These vital statistics should include the employer's name, address, home and work phone numbers, the children's names and ages, the family doctor's name and phone number, and any known allergies or special needs. You may also want to list some of the children's favorite foods, toys, and games.
3. If at all possible, plan to visit an employer's home before you baby-sit for the first time. This will let you get to know the children and pets. You should also ask to be shown the location of first-aid supplies and the master electrical switch. You'll want to know

how to handle the burglar alarm system, too, if there is one, and any appliances you'll be expected to use.

4. Assemble a sitter's bag to take with you when you go to baby-sit. This should contain a flashlight, in case there's a power failure in a strange house; Band-Aids and antibacterial spray or cream for quick aid if a child gets a cut or scrape; a favorite picture book and a favorite book to read to older children; a few balloons; and a game to share.
5. Don't forget to be concerned for your own safety, too. If the doorbell rings while you're baby-sitting and you haven't been told to expect someone, you don't have to answer. And while it's important to answer the phone in case the parents might be calling, you should hang up immediately on an obscene caller. Plan with your parents in advance how you should get home if your employer returns too drunk to drive. You might carry cab fare. Or your parents may prefer you call them for a ride.

Like other jobs, training is available for this job. You may want to check with the American Red Cross or churches in your community.

What's inside a tree bud? Why do flowers smell sweet? You'll find out and discover a lot more about plant life in springtime—next.

4. BUDS AND BLOOMS

From Flowers to Seeds

WHERE WINTERS are warm, flowering plants may bloom year-round. But where winters are cold, spring is the season when many flowering plants burst into bloom. The flower's job is to make seeds. Seeds—like the animal eggs you investigated in the previous chapter—are life packages. They contain an embryo or young plant and a food supply inside a protective coat.

You'll remember that baby animals usually start life when a sperm cell and an ovum fuse. Plant embryos are also usually created when male and female plant cells join. The male cells are called pollen. They're formed in the anther, the top part of the stamen, or male part of the flower. Pollen looks like a powdery dust when it's released. Most pollen is yellow, although the pollen for some kinds of flowers is black, white, red, or even blue. And, when viewed under a microscope, each type of pollen has its own distinctive shape.

daisy   *Kentucky bluegrass*   *coleus*

The female cells are called ovules. They're attached to the walls of the ovary or female part of the flower. To produce a new, young plant, a pollen grain from the same kind of plant must enter an ovule. The ovules can't move. So each correct kind of pollen grain that lands on the stigma, the sticky top of the ovary, goes down to an ovule.

Next, the ovary develops into a protective covering for the seeds.

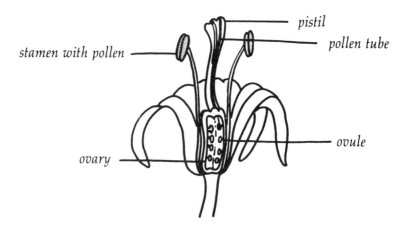

It may form a hard shell like an acorn, or it may develop a papery wing like a maple seed. And there may be additional fleshy parts to provide even more protection. For example, the core of an apple is the ovary, while the part you eat is formed from a fleshy flower stalk called a receptacle that swells around the ovary.

RIDDLE

What flowers do all people have?
Tulips.

Flower Helpers

PLANTS CAN'T go hunting for a mate the way animals can. Plants can only stay in one place. Some plants, such as tomatoes, have flowers with both male and female parts. Then the pollen can simply fall on the stigma. So these flowers are able to pollinate themselves. Other plants, though, such as corn, have flowers with only male parts or only female parts. These flowers must depend on a helper to transport pollen from the male to the female part.

The wind is one helper. Flowers that depend on the wind produce millions of pollen grains. Then even though much of the pollen may be lost, some is likely to fall on flowers of the same kind of plant. Watch for yellow dust coating cars or floating on puddles in the spring. That's pollen. Oak trees, grasses, and ragweed are among the plants pollinated by the wind.

Insects may be the best helpers. Butterflies, moths, certain wasps and beetles, and especially bees transport pollen. But insects don't help on purpose. They visit the flowers to eat nectar—a sticky, sweet liquid—and pollen, which is rich in protein, fats, minerals, and vitamins. Honeybees are particularly good helpers because they tend to visit whole groups of the same kind of flower. Then the pollen that sticks to their stiff body hairs is likely to rub off on another flower of that same kind of plant.

Have you ever gotten a "mustache" drinking milk? When bats poke their long noses into flowers to drink the sticky, sweet nectar, they get a pollen mustache. Some of the pollen rubs off on the next flower the bat sips from.

A number of birds sip nectar and so transport pollen, too. The one you're most likely to spot, though, is a hummingbird. This little bird,

Each hind leg has a kind of comb the bee uses to collect the sticky pollen.

not much bigger than a bumblebee, carries pollen on its head feathers.

Besides producing nectar that the insects and animals want, flowers that depend on these helpers are especially designed to be sure their pollen will be picked up. First, these flowers are often bright colors or produce scents to attract visitors. Then the nectar is likely to be tucked deep inside. So the helper must brush past the male or female flower parts to obtain its reward. And to be sure the visitor will know where to find the nectar, the flower often has a bright pattern that points the way. Some flowers have triggers that, when touched, shower the visitor with pollen. Producing new life is serious business, so nothing is left to chance.

Smell Hunt

ON A warm, sunny spring day when there isn't much wind, go outdoors, close your eyes, and sniff deeply. Then count the number of distinctly different scents you smell. Can you follow one scent to a flower?

Ha-a-a-a Choo!

SPRING ISN'T a happy time for everyone. For some—maybe you—it's a time of sneezing, itchy eyes, and a runny nose. Those people are allergic to the pollen that plants release in great quantities in the spring. Pollen isn't the only cause of allergies. People may be sensitive to bee stings, certain foods, pets—almost anything. But the most common cause of an allergic reaction is pollen.  This is also the hardest substance to avoid. People allergic to dogs can simply avoid dogs. And people who are affected by chocolate don't have to eat chocolate. But pollen is microscopic and floats in the air. So pollen may be entering your body with every breath you take.

If you're allergic to pollen, your body responds to these plant cells as if they were harmful invaders. Your body's immune system immediately goes into action, releasing special chemicals designed to fight off the pollen. It's these chemicals that cause you to sneeze, develop red itchy welts, and make your eyes water. If you are among the fifteen million Americans sensitive to pollen, you'll be glad to know that air conditioners help. As the air is cooled, it's also filtered, getting rid of most of the pollen grains. Rain helps clean the air, too.

If you're especially sensitive to pollen, though, you may need to visit an allergist, a doctor who specializes in such problems. Although they're not sure why it happens, allergists have discovered that a series of injections of extracts of the problem-causing substance usually helps the person build up a tolerance. You may not like the idea of taking a series of shots, but in the long run, this treatment could let you breathe easier.

Wildflowers

GARDEN FLOWERS are just wildflowers that have been carefully tended, with only the best plants selected for replanting over many years. Orchids, for example, are related to the wildflowers called lady's slippers. While all flowers are a pleasure, wildflowers are a special treat because discovering them is either unexpected or the result of a diligent search. Here are some wildflowers you can hunt for in the spring. To see what they really look like, place a sheet of tracing paper over the page, trace the flower, and then color it according to the directions.

You could put your wildflowers on index cards and use them to test how many you can identify on sight. Or you might start a collection of wildflower pictures, adding on drawings or photos of other wildflowers you discover. Pictures are a nice way to enjoy the flowers and still leave the real plants growing in the wild. Make a note on each card about where you spotted that flower and the date you saw it blooming. Then you'll be able to find it again next spring.

Hepatica
(also called Liverwort)

THIS IS one of the first wildflowers of spring. You'll find it in wooded areas throughout the United States. But you'll need to look close to the ground. The leaves are only three to four inches above the surface, and the flowers are on a six-inch stalk. The liverwort got its name because its leaves have three lobes just like the human liver. During the Middle Ages people thought this plant might provide a cure for liver diseases.

The flowers may be blue, pink, or white. The leaves are brownish green.

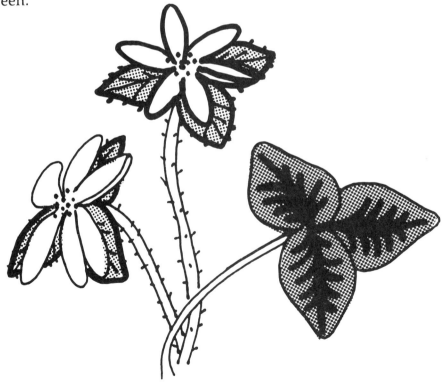

Bloodroot

THE FLOWER first appears wrapped in its own leaf, as if cloaked against lingering cold weather. And although the flower is beautiful, it may be the large, thick leaf standing on an eight-inch tall stem that first catches your attention. You'll find it in most woods east of the Rocky Mountains. Count yourself lucky if you find the bloodroot in full bloom. The flowers fade quickly. Sound like a gruesome name for such a pretty wildflower? This plant's root gives off a bright red juice that the American Indians are said to have used both to dye their clothes and to make war paint.

The flowers are white with yellow centers. The leaves are dark green.

RIDDLE

What letter is an insect?

B

Longspur Violet

THIS INTERESTING-LOOKING violet is only one of nearly a hundred varieties of this hardy plant. You'll find violets throughout the United States wherever the soil stays moist. While many are violet colored, some are yellow or white. And one variety of violet has green flowers. If you find a wild violet, look at this ground-hugging plant closely. Most varieties produce two types of flowers. You'll easily spot the showy blooms. These depend on bees and butterflies attracted by their sweet scent to pollinate them. But there are usually also small, hard-to-see white flowers below the leaves. These blooms are capable of self-pollination, guaranteeing that the violet will be able to reproduce.

The flowers are purple with yellow centers. The leaves are bright green.

Red Trillium

IT'S HARD to miss this wildflower in the woods. This big plant—often over a foot tall—grows throughout the United States. If the red, bell-shaped flower doesn't help you identify it, it's three large leaves will. In fact, all the parts of this plant—petals, sepals, and leaves—are in groups of threes. That's why it's called "trillium" after the Latin word for three. If you don't see the red flower at first, peek under the leaves. These pretty flowers sometimes hide. Trilliums are also called wake-robins because they appear in northern woods about the same time as the robins.

The flowers are red, the sepals are bright green, and the leaves are dark green.

Pink Lady's Slipper

CAN YOU guess why the lady's slipper is sometimes called the moccasin flower? It grows in moist woods throughout the United States—except California and Florida. Many of the country's moist woodlands have been cleared, and in the places where they still grow, people have picked this beautiful plant almost to extinction. Picking this fragile flower is now against the law in some states. But even if it's not illegal to pick it where you live, leave the lady's slipper for others to enjoy, too.

The flower is a dark rosy pink and the leaves are bright green.

Jack-in-the-Pulpit

THIS UNUSUAL flower is a treat to find. It blooms from April to June in moist woods as far south as South Carolina and as far west as Kansas. Watch for the cobralike hood atop a two- or even a three-foot stalk. The leaves are divided into three leaflets and are even taller than the hooded flower. Pay special attention to where you find the jack-in-the-pulpit because you'll want to look at this plant again during the summer. By then the flower will have fallen away, but the stalk will be topped with a cluster of bright red fruit. This is poisonous, though, so just look.

The lines on the flower are yellow-green. The outside of the bloom is bright green and the inside is purplish brown. The stalk is also purplish brown. The leaves are bright green.

Whistling Dixie

YOU PROBABLY won't be able to play that tune, but you'll definitely produce some interesting sounds with a grass whistle. First, find a long, flat blade of grass. Break it off and hold it, stretched

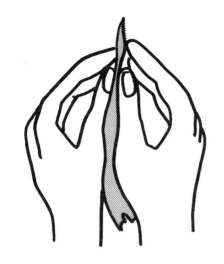

tightly by your palms and your fingers, between your thumbs. Then put your lips against your thumbs and blow hard. The blade of grass will vibrate, making the air vibrate and produce a sound. Does changing how tightly the grass blade is stretched affect the whistle's sound? Do different blades of grass produce different sounds?

Sap's Rising

IN THE spring, stimulated by special chemicals called hormones, much of a tree's supply of stored starch is changed into sugar. Then water is drawn in through the roots. This creates a pressure that drives the sugar-rich sap upward, supplying energy for the new leaves and blooms. But you've seen wood. It appears to be solid. So how does this sap move up through the tree—like a sponge soaking up water or through special tubes?

You can do an experiment to find out. A tree is too big to work with, but luckily liquids flow through celery just as they do through a tree. So pick a celery stalk that has leaves. And hold the base underwater in a pan while you snip off about an inch with kitchen shears. This lets you expose plant tissue without letting air, which could block the cells, get inside. Next, pour a half-cup of water into a tall glass, add several drops of red food coloring, and quickly transfer the celery stalk to the glass. Let it sit overnight.

When you take the celery stalk out of the water, examine the base. The little dots you see are the ends of plant tissue, called phloem, that forms tubes through the stalk. To prove that the colored water moved up through these tubes to reach the leaves, carefully snip the stalk open lengthwise. You'll be able to follow a red phloem tube all the way up.

In trees, the phloem layer is very close to the bark. So holes are only drilled a half-inch deep into a sugar maple's trunk to tap it for sap. Although you may think of summer and fall as harvest times, this sap is actually the first harvest of the year. It's collected from the trees—usually in March—as soon as the temperature begins to rise to around 40°F during the day. Then the sweet sap is boiled to produce maple syrup.

Surprise Package

THAT'S WHAT each spring bud is. To find out what's inside, pick a swelling bud from a tree or bush and carefully open it. Then use a pin to separate any tiny leaves you find inside. How many leaves are there? Compare this number with what you find inside a bud from a different tree or bush. Don't pick more than one bud from each plant, though. These leaves are needed to produce food. Large buds, such as those you'll find on horse chestnut trees, contain an extra surprise—flowers.

Blooming Trees

FLOWERS AREN'T the only springtime bloomers. Many trees bloom, too. While some trees are cultivated just for their flowery display, many others produce blooms that go unnoticed. This is particularly true of trees, such as oaks, that depend on the wind for pollination. Their blooms don't have to attract a helper.

Start watching for tree flowers early in March. You'll be treated to some blooms you may have been missing.

Pussy Willow

THOSE FAMILIAR furry puffs you find on this plant are blooms just starting to open. What you can't tell at this stage, though, is that pussy willows actually produce separate male and female flowers on different plants. For this reason, pussy willows depend on insects to transfer their pollen. So at the base of each furry bloom, there is a little pocket of nectar.

female flowers   *male flowers*

Would you like an early spring treat? Once the temperature has been above 37°F for at least four weeks, you can force pussy willows into bloom. Cut three or four branches from a tree, bring them indoors, and put the cut ends into a vase half-full of water. Then put the vase in a sunny place. Within a couple of days, the pussy willow branches will be covered with fuzzy blooms.

Horse Chestnut

THIS TREE'S flowers are spectacular spikes of blooms that open in May. The nectar's sweet scent attracts bees and wasps. But even with lots of insect activity, not every one of the hundreds of blooms will be pollinated. The seeds that do form become large nuts inside green prickly husks. This nut is poisonous, though, so don't try to eat it.

Sugar Maple

AS SOON as you see leaves opening on this tree, look for the blooms. Male and female flowers will appear on the same branches. But they'll be in separate clumps. You can recognize the male flowers by their yellow stamens. Tie a colored strip of cloth around a branch near a cluster of female blooms. Then continue to watch this spot through the spring and summer. After the ovules are fertilized, they change slowly. The ovaries swell and develop wings. The mature winged seeds or samaras will spin away from the tree like little helicopters in the wind.

Lettuce Eat

THE AVERAGE American eats about twenty-one pounds of lettuce a year. Are you eating your share? This vegetable is a good source of fiber—needed by the body for good digestion. And it's been a popular vegetable for centuries. The Romans, in fact, were so fond of one type of lettuce called cos that today this kind is known as romaine lettuce. And it's said that Augustus Caesar, believing that eating lettuce had cured him of an illness, even built a shrine honoring this vegetable. Christopher Columbus helped bring lettuce to the New World. It was grown in California by the Spanish priests at their missions, and today—thanks to improved methods used to keep the lettuce fresh during shipping—California is the leading supplier of lettuce to the United States year-round.

There are four basic types of lettuce: crisp head (also called iceberg) with outer green leaves and white inner leaves; butter head, which looks like a small cabbage; loose head, with open leaves; and cos, which grows as upright leaves rather than as a head. There are a number of different varieties of each of these four basic lettuce types. Some can even withstand the summer's heat. Most varieties of lettuce, though, prefer cool weather. So this is a crop you can plant early to eat in the spring.

To raise your own lettuce, first buy seeds for the variety you like best. Then plant about a dozen—one to a peat pot (available at garden-supply stores). Put the pots on a tray, water as needed to keep them moist, and set the tray in a warm, sunny spot indoors. When the seedlings are about an inch tall, transplant them to your garden. Or grow them in six-inch deep flowerpots—one plant to a pot. (Loose head varieties do well in pots.) To be sure the pot has good drainage, you may want to cover the bottom with gravel before you fill it with potting soil. Water as needed to keep the soil from completely drying out, and in about fifty days you'll have fresh lettuce for your salad.

Pocket Salad

THERE'S NO end to the tasty ingredients you can add to salad. But here's a hardy lettuce meal you can tuck into an edible package and carry with you. You'll need a pocket bread (available at the deli section of grocery stores). Next, rinse and tear up about a half-cup of lettuce. Then wash several small cherry tomatoes, collect one-fourth cup of grated cheese (your favorite kind), and one-fourth cup of diced luncheon meat (your favorite variety). Carefully slit an opening into the bread. Mix the lettuce with the other ingredients as you pack them into the pocket. If you're not ready to eat this immediately, slip the whole thing into a self-sealing plastic bag.

Crazy about Tulips

THIS FLOWER, often considered a symbol of spring, is believed to have its beginnings in ancient Turkey or Central Asia. But it was in Holland that the tulip first became recognized as something very special.

In fact, from 1634 through 1637, the Dutch were caught up in what has been called Tulipomania because they were just crazy about tulips.

The excitement started when

someone discovered that unlike other flowers, tulips have the habit of "breaking" sooner or later. Breaking is the term used to describe how tulip petals may be streaked with other colors. Once a tulip develops a new pattern, this change is permanent. So the seeds of the tulip with this new pattern can be used to produce more tulips like it.

People were willing to pay such a high price for unusual tulips that raising tulips became a fad and bulbs were planted anywhere there was a bit of land. The hope was always that one tulip would break, producing a spectacular new and valuable bloom. Some people even invested in other people's tulips—just like someone might invest in a mining or drilling venture, hoping to strike it rich. And, as you might guess, the price of bulbs skyrocketed. On one occasion, for example, a single treasured tulip bulb was the only price asked by someone selling a well-established business. And one prized bulb was considered enough to offer as a wealthy bride's dowery.

Then, as quickly as the interest in tulips had begun, it was over. Tired of bargaining over tulips, everyone began to sell their bulbs. And since the market was flooded with tulips, the value of the bulbs plummeted. People who had invested all their money in tulips suddenly found themselves owning nearly worthless property. There was panic and ruin. Finally, in an effort to stabilize the economy, the government seized control and began regulating the tulip trade.

Luckily, tulip bulbs have remained fairly inexpensive. So if you'd like to enjoy some of these flowers in the spring, you'll just need to plan ahead. Tulip bulbs need to be planted in the late summer or fall, except in very warm climates, where they may be planted in late fall or winter. The plant needs enough time for the roots to develop before the ground freezes. Tulip seeds aren't planted because plants started this way don't bloom for several years. Buy some bulbs and follow the planting directions on the package. Then watch for these plants to poke up in the spring.

Is the wind blowing hard? It often does in the spring. You'll discover why in the next chapter. And you'll find some activities perfect for wild and windy springtime weather—next.

5. WILD AND WINDY

Why Is March Windy?

THERE ARE two kinds of winds: local winds that only affect small areas and global winds that affect large areas. And both tend to blow harder in the spring, particularly during March, because of the sun.

Wonder what the sun has to do with how hard the wind blows? To find out, first observe what happens in this activity. You'll need an empty glass bottle with a small opening, an electric skillet (or a saucepan on a stove), water, and a rubber balloon. Put the balloon over the mouth of the bottle. Pour enough water into the skillet to cover the bottom. Then put the bottle in the center of the skillet and switch it on to medium high. Heat for five minutes, adding more water as needed to keep the skillet from going dry.

As the air inside the bottle heats up, you'll see the balloon lift and swell. Without the balloon, the invisible air inside the bottle would simply have escaped, and you wouldn't have been able to tell that it was rising. As added proof that the air rose because it was heated, switch off the skillet and let the bottle cool. Soon you'll see the balloon shrink and finally collapse as the air cools and sinks back into the bottle.

What happened to the air trapped inside the bottle happens to air everywhere. As the earth's air is warmed, it rises. And as quickly as the warm air moves up higher into the atmosphere, cool air rolls in under it to fill in the space. Moving air is wind.

How strong the wind blows, then, depends on how great the temperature difference is between the areas of warm and cool air. And spring is a time when the air over some areas is suddenly getting much warmer, while the air over other areas remains ice cold.

Clouds also help trap radiated heat energy.

Just having the sun shine isn't enough to heat the air up, though. The sun's energy has very little effect on the air as it passes through it. Instead, the sun's energy has to be absorbed by matter—rocks, soil, water, cars, houses—almost anything. Then the matter heats up and radiates the heat. It's this radiated warmth that heats the air.

To prove that different kinds of matter heat up and radiate heat differently, try this test. You'll need an indoor/outdoor thermometer, a watch, a ruler, a notebook, and a pencil. On a bright sunny day when there is very little wind, go on a matter hunt. Each time you find a kind of matter you want to test, such as an area of bare soil, write it down in your notebook. Next, hold the thermometer suspended a foot above the surface for three minutes. Then check the temperature and record that next to the test matter's name.

Which kind of matter radiates the most heat? Do darker materials seem to heat up more or less than lighter ones? Why do you think it was important to conduct this test when the wind was calm?

How Windy Is It?

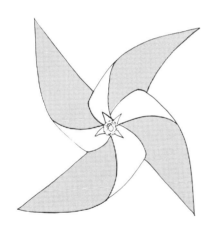

ONCE YOU'VE discovered why spring is so windy, you can make a pinwheel to test the wind's strength. The stronger the wind the faster the pinwheel will spin. To make a pinwheel, you'll need: a sturdy yet bendable six-inch square of vinyl or plastic (could be cut from the kind of clear vinyl binders sold for reports), one plastic straw, a twelve-inch-long pipe cleaner (sold in stores carrying craft supplies), scissors, and a hole punch.

First, you'll need to create the pinwheel's basic shape. So carefully fold the vinyl square diagonally. Crease it. Then open it and fold it diagonally again in the opposite direction, leaving an X of creases. Next, use the scissors to cut these crease lines from the corners toward the center. Stop about an inch from the center. Then make a hole in the right corner of each of the triangles. And punch a hole in the very center, too. Be sure these holes are just large enough for the thin straw to slip through easily.

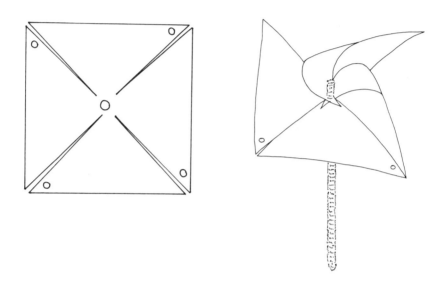

You can now form the pinwheel's wind vanes. First, push the pipe cleaner up through the center hole so that about two inches sticks up above the vinyl square. Next, bend one triangle over, threading the pipe cleaner through its hole. And do the same with each of the other triangles, moving in counterclockwise order around the pinwheel. When you've finished, hold the pinwheel vanes tightly together and bend the end of the pipe cleaner sticking up through the center to form a circle about a half inch in diameter. Keep twisting if necessary to make this a block the vanes won't slide off. Then slide the straw onto the pipe cleaner and through the center of the pinwheel vanes until it touches the blocking circle. Twist the free end of the pipe cleaner to keep the straw from slipping back. Finally, let the pinwheel vanes slide back onto the straw.

To try out your pinwheel, go outdoors and hold it with the vanes facing into the wind. The wind will be full in your face if you're facing into it. Count how many times the pinwheel goes around in fifteen seconds and multiply that by four to see how fast it rotates per minute. You may want to stick a piece of brightly colored tape on one vane or color one vane with a marking pen to make it easier to tell one rotation. If the wind is gusting, conduct three tests, adding up the total and dividing by three to compute the average. Use your pinwheel to test the wind in different locations, such as at the base of a tree, in an open area, and close to a building. Does the wind blow harder in one place than it does in another?

You may want to make a chart and use your pinwheel to test the wind's strength at the same time each day for a week. Or you could check every hour throughout one day.

RIDDLE

What did the rug say to the floor?
Don't move. I've got you covered.

Flights of Fancy

MARCH WINDS are perfect for flying kites. Whether you follow the directions provided in the Kite Builder's Kit on page 94 to build your own kite or purchase one, it helps to understand why kites fly.

Because you're holding a line attached to one end, the surface of the kite face is held at an angle to the wind. This causes air to flow down the kite's face. As the air molecules reach the base of the kite, they escape and slip rapidly up the back side of the kite. Like traffic building up on the freeway, though, the molecules don't escape fast enough. And air molecules pile up at the kite's base. The more molecules pile up, the more the air moving over the kite face is slowed down. Slow-moving air has greater pressure than fast-moving air. So the air pushing up against the kite's face has more pressure than the air pushing down on the back. This makes the kite hover. If the pressure below becomes great enough, the kite rises.

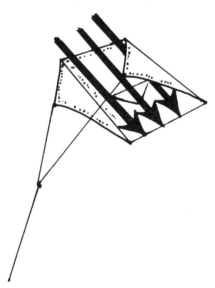

If you have trouble believing that fast-moving air has less pressure than slow-moving air, try this activity. Cut a two-inch wide strip down the whole length of a sheet of notebook paper. Next, hold one end of this strip stretched between your hands just beneath your mouth. Take a deep breath and blow down hard across the top of the paper strip. Surprise! Your fast-  moving air doesn't force the strip down. Instead, like a kite, the paper rises when there's fast-moving air above and slow-moving air below.

If you find the hardest part of flying a kite is getting it airborne, try this tip. First, check that there's some wind blowing but that the wind isn't blowing too hard. An ideal kite-flying wind is between five and fifteen miles per hour. If you aren't sure of the wind speed, look for some natural signs. Leaves should be moving. It's also a good sign to see dust blowing or flags lifting. But if small trees are swaying or flags are flapping hard, the wind's probably too strong. Never run with your kite. It there's enough wind, you don't need to run. When you're running, you can't watch where you're going as well as you should. You could stumble and fall.

Instead, stand with your back to the wind while a friend carries your kite away, holding the top pointed straight up. When about twenty-five feet appear to separate the two of you, call for your friend to release the kite as you begin to pull in on the line. The kite should climb immediately. If it doesn't, repeat this process.

Once your kite is airborne, you can help it rise higher by using a technique called "pumping." First, pull in on the line a little. You'll immediately feel the kite try to tug the line out of your hand. So feed out more line until you feel the line go slack in your hand. Each time you repeat this pumping process, the kite will soar a little higher.

Many communities hold kite festivals. Check if there's one in your area. You may want to participate. If you lived in Japan, you'd see lots of kites flying on the fifth of May. That date is celebrated as "Boys' Festival" and kites are traditionally flown in honor of good sons.

Kite Builder's Kit

THE SLED kite or Scott sled (named after its inventor, Frank Scott) is one of the easiest kites to build and fly. You'll need three thirty-six inch wooden dowels that are one-eighth inch in diameter, strong tape (such as electrician's tape), transparent tape, a hole punch, scissors, a yardstick, a large polyethylene trash bag, and a roll of monofilament fishing line.

First, cut the kite's cover out of the plastic trash bag, following the dimensions on the diagram. Don't forget to cut out the vent. This

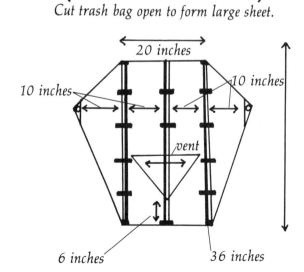

helps make the kite a more stable flyer. Next, use strong tape to anchor the three wooden spines as shown. Then, reinforce the two wingtips with transparent tape and punch a small hole in each corner. Finally, attach a piece of the fishing line seventy-two inches long to each of these holes. And tie the two free ends together. This forms the bridle, designed to let the kite be angled against the wind.

To fly your kite, attach the bridle to the roll of fishing line. A stick poked through the open center of the roll will let you feed out the line more easily.

> *RIDDLE*
>
> *What would you get if you crossed a quarter-pound of ground beef and a bumblebee?*
>
> *A humbrger.* (upside down)

Kite Flyer's Safety Tips

TO HAVE the most fun flying your kite, you'll want to protect yourself and your kite. Here are some ways to be sure you do both.

1. Always fly your kite in an open area away from wires.
2. Don't fly too close to other kites.
3. Never fly your kite during a thunderstorm.
4. Don't let your kite go too high. The legal ceiling for kite flying is five hundred feet.

Play It Again, Wind

ANOTHER WAY to enjoy the strong spring winds is to make a wind chime. You'll need a large, sturdy plastic lid (the kind that comes on large plastic butter tubs and plastic ice cream buckets), monofilament fishing line, a hammer, a nail, a large sewing needle, electrician's tape, and scissors. This will form the hanging frame for your wind chime.

You'll also need to collect whatever you want to hang from it. These should be items that can be easily tied to a line, that will make a pretty sound when they bump into each other, and that will be sturdy enough not to break in the wind. You could use metal buttons, sections of hollow bamboo, or individual-sized metal juice cans. You'll need enough so that the chimes are almost touching when they're suspended. The example below uses small metal juice cans.

Start by using the needle to poke two small holes about a half-inch apart in the center of the plastic lid. Cut a two-foot piece of fishing line, thread it through these holes, and tie the two ends together in a knot. This loop is what you'll use to hang up your finished wind chime.

Next, plan how far apart you'll need to hang your chimes. Poke a hole through the plastic lid and cut a one-foot-long line for each chime. Thread the line through the chime before attaching it to the lid. You'll need to use the nail and hammer to punch a small hole in the bottom of each juice can. You may want an adult to help you with this job. To be sure the line won't pull through, tie the end that will be inside the can to a button. Next thread the line

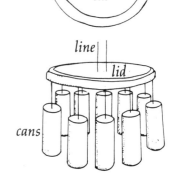

through the hole in the can. Then thread it through the lid from the bottom side, tie a knot on top of the plastic lid, and secure it with a strip of tape. Hang your finished chime in the open near a window or on a porch where you can listen to its music anytime the wind blows.

The Storm

IT'S BEEN a warm, sunny afternoon. But now you notice that the wind is blowing harder. A few flower petals whip by. You rub your arms, feel goose bumps, and wish you had a jacket. It's getting cooler fast.

Towering white skyscraper clouds are billowing even higher. The bottom edge of these clouds are darkening. First they look gray. Then they appear charcoal.

The air even smells different now. Small trees are swaying. Bigger trees are rattling their branches. A flag is flapping against its flagpole.

Suddenly, in the distance, you see a flash of lightning. And after a little bit, thunder rumbles. Then a big raindrop plops on your arm. Another lands on your nose. Lightning flashes again, and this time the thunder follows more quickly. The raindrops begin to fall faster and faster until the wet spots on your shirt run into each other.

You need to find shelter. But you aren't the only one that needs to stay dry and be safe. Where do you think the birds might go during the storm? The ants? The spiders? The squirrels?

Sounds Like Thunder

SPRING STORMS are often accompanied by lightning and loud booms of thunder. But you don't have to wait for a storm to hear thunder. You can produce your own. All you'll need is a Boomer. To make one, collect a twelve-inch square of sturdy poster board, a single-edge razor blade, a ruler, a pencil, a large brown paper bag, and package-sealing tape. Then follow these directions:

1. Lay the ruler diagonally across the poster board square and draw a line connecting two of the corners. Then slice along this line with the razor blade so the poster board will bend easily. Be careful not to cut all the way through.

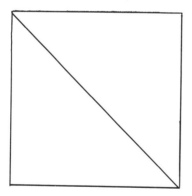
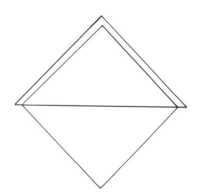

2. Cut open the paper bag to form one big sheet of brown paper. Fold the poster board along the line you sliced, creating a triangle. Place the poster board triangle near the middle of the brown paper, at least an inch from the corner. Then measure out a triangle one inch larger on each side than the poster board triangle. Cut out the paper bag triangle.

3. Open the poster board with the sliced side up and arrange it so it looks like a diamond. Lay the paper triangle over the poster board with the bottom edge across the center of the diamond. Fold the overlapping edge of the paper triangle to the back of the poster board diamond and tape it. Fold the two sides of the poster board diamond together—brown paper inside.

To create a boom of thunder, pinch the cardboard points together in one hand and hold your arm up about head high. Then jerk your arm down hard and fast. If you don't hear a loud boom, tuck the paper triangle inside and try again, jerking your arm down harder and faster.

When you snap the Boomer, you're forcing air to move quickly. And the sudden vibrations created by this rush produce the thunderlike sound. In a springtime thunderstorm, lightning makes the air move by suddenly heating it. The results are the same—a noisy bang!

A Portable Roof

NO ONE's sure who first came up with the idea for the umbrella. And while it's considered mainly a rain shelter today, it was probably first used as portable shade. In fact, the name umbrella refers to the fact that it creates a shadow—just right for getting out of the hot sun.

The ancient Egyptians considered the umbrella a heavenly symbol. So standing under one was a privilege reserved for the king. In fact, for centuries, only the wealthy had umbrellas. It wasn't until the 1600s that Europeans began to consider umbrellas good protection against rainy weather. The early European umbrellas were heavy, awkward structures with sturdy oak handles and whalebone ribs supporting the cloth. These umbrellas also required special care. If the whale ribs got too wet, they lost their elasticity and collapsed. If they weren't carefully dried, the ribs became brittle and were likely to snap unexpectedly. And even if the umbrella stayed properly open, the covering was only oiled cloth. So once it became water-soaked, it often leaked.

By the late 1700s, though, the umbrella had become a hot fashion item as well as something useful. Then umbrella manufacturers competed for business by including special extras. Some umbrellas had hollowed out handles containing perfume bottles, powder puffs, and other items. For those wanting total protection, a dagger could be pulled from the handle. Other models of umbrellas had built-in telescopes and windows. For people who were afraid of being caught in a storm, there was even an umbrella with a lightning rod on top and a long trailing ground wire.

What do today's umbrellas have to offer? Visit a store that sells umbrellas. Check what different sizes are available, what materials the umbrellas are made of, and what special features, if any, these umbrellas provide. Then think about designing your very own custom-made umbrella. What special features would you include?

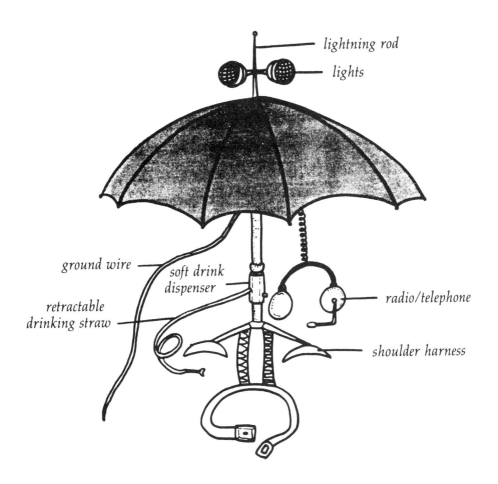

Give Me a Sign

INSTEAD OF carrying an umbrella wherever you go, it would help to know if it's likely to rain. Watching the clouds can help you predict the weather. Meteorologists have observed clouds and weather conditions over many years, and they've discovered that certain types of clouds provide pretty reliable forecasts.

Cirrus: These very high clouds are made of ice crystals. Because they form when a warm air mass slides up over a colder one, cirrus clouds may signal a storm is coming.

Cumulus: Looking like balls of cotton, cumulus clouds form when one area, such as a parking lot, becomes warmer than the land around it. When the air rising from this spot gets high enough, the water vapor in the air condenses and produces a white puff. Cumulus clouds are fair-weather clouds.

Stratus: The whole sky may look hazy. If the clouds begin to turn dark, they're called nimbostratus clouds. Then a slow, steady rain is likely.

Cumulonimbus: Head for cover when you see these dark, billowing cloud monsters. They produce thunderstorms and even tornadoes.

If you really want to see a lot of cumulonimbus clouds, visit the island of Java in Indonesia. That area averages three hundred twenty-two days of thunderstorms a year.

> RIDDLE
> What did the bee say to the rose?
> *¡pnq 'ıH*

How Rainy Is April?

ACCORDING TO the saying, it's this month's rain that brings May flowers. But is this really the rainiest spring month? To find out, you'll need to collect rain all spring—from March to June. And to do that you'll need a two-liter plastic soft-drink bottle with the top cut off. Or use an empty three pound can. Place this outdoors in the open.

Check after each rain, and if you've collected any water, stick a ruler straight down to measure the depth. An inch is a lot of rain, so it will probably be less than an inch. Make a chart like this one to keep track of the rainfall in your area. You may want to compare how much it rains at your house with the official amount reported by your local television or radio station.

Date	Inches of Rain

Did April turn out to be true to its reputation as the rainiest month? You may want to keep on collecting for a whole year to see if spring really is the rainiest season.

Every season seems to have its own special holidays. Spring does, too. You won't want to miss any of the events coming up in the next chapter.

6.
STRICTLY SEASONAL

April Fool

THE FIRST of April is traditionally a day for playing practical jokes. And while this unofficial holiday is observed in countries as far apart as Sweden and India, no one is quite sure of it's origin. One explanation is that it comes near the spring equinox (usually March 21), when the weather plays tricks, switching from sunshine to rain and back to sunshine unexpectedly.

Some animals play tricks, too, but they're not joking. This is what they do every day to stay alive.

Hiding

SOMETIMES THE trick is not to be seen. And some animals are experts. One type of moth, commonly called the underwing moth, has brightly colored lower wings. But when this moth wants to avoid becoming a bird's dinner, it lands on a tree trunk and tucks its lower wings under its upper wings. Then it looks just like a piece of bark. Curiously, near cities, where factory soot has caused tree trunks to become darker, only dark-colored underwings have survived. So now these moths generally have much darker wings than their ancestors. Even some with totally black wings have been observed.

Flower mantids blend in beautifully. These hunters are able to ambush insects visiting orchids because their body parts are pink and petal-shaped.

On the other hand, you'd think everybody would see right through the dragonfly's disguise. But that's the whole idea. This insect's big wings are transparent. So it has no trouble sneaking up on its prey unnoticed.

Not What It Seems

ANIMALS WHO are likely to end up as someone else's dinner often come disguised as something not worth eating. An inch worm, also called a stick caterpillar, for example, reacts to an approaching bird by arching its body away from the branch that it's crawling along. Then, stretched out and stiff, the little worm looks like just another branch rather than a tasty snack.

Thorn bugs look like thorns, dead leaf butterflies look like dead leaves, and box turtles look like rocks. But the slug caterpillar's disguise deserves the grand prize. When this little creature flattens out on a leaf, it looks exactly like bird droppings.

The anglerfish, on the other hand, depends on a trick to catch its dinner. This fish looks a lot like a chunk of rock when it's sitting quietly on the ocean bottom. But just above the anglerfish's mouth there's a little polelike fin with a frilly end. When it waves this lure, curious little fish swim closer. Then gulp! Dinner is served.

Mistaken Identity

SOMETIMES IT pays to have others think you're somebody you're not. Birds, for example, seldom try to eat a monarch butterfly more than once. They seem to think those butterflies taste awful. So viceroy butterflies, which look a lot like monarchs, manage to escape, too.

Bumblebee moths use this same trick. They can't sting the way a bumblebee can, but they look like bumblebees. Frogs and birds don't seem to want to take a chance, because they leave both the real bee and the imitation bee alone.

The io moth takes a slightly different approach. When danger threatens, it spreads out its lower wings, exposing big bright eye spots. With any luck, the enemy gets scared and bugs off.

RIDDLE

Why should you never tell secrets to pigs?
Because they're squealers.

Go A-Maying

IN ANCIENT times, May first was one of the most important holidays of the year. And because it came at a time when winter was so obviously giving way to new life, May Day was a celebration of nature.

Even before the sun was up on this special day, women—both nobility and peasants—rushed to the fields. According to tradition, anyone who washed their face with dew on that particular morning could count on becoming more beautiful.

By the time the women had clean faces, everyone was outdoors collecting flowers and cutting tree boughs for the maypole. The maypole was a tall, bare tree trunk set up on the village green. In some cities, a particularly tall and sturdy tree was permanently erected for this event. London in the late 1600s, for example, was reported to have a cedar maypole one hundred thirty-four feet tall. But whatever its height, on May Day, this bare tree bloomed as garlands of flowers and colorful streamers were attached to it. Then dancers, holding on to the free ends of the streamers, wove them into intricate patterns as they danced around and around the pole.

Another special May Day tradition was giving May baskets. The night before, small baskets were filled with flowers and hung on the doorknobs of friends and neighbors. If this is a May Day tradition you'd like to continue, follow the directions on the next page to make a May basket. Instead of flowers, you could put a special message or a treat in the basket. Think what a nice surprise this will make for May Day morning!

To make a May basket, you'll need an 8½ x 11 inch sheet of paper (white or colored), scissors, and tape.

1. Make the paper into a square by lifting a lower corner and pulling

it across to the opposite edge. Crease lightly. Then trim off the excess strip.
2. Fold the paper square carefully into thirds. Open and fold it in thirds again in the opposite direction.
3. Cut in along the bottom edge of squares 1 and 2 to the corner of the middle square. Then cut along the top edge of squares 3 and 4 to the corner of the middle square. If you want to, you can decorate the basket.
4. Fold the two outside squares halfway across the middle square and tape to secure them from the inside. Repeat with the opposite square.
5. Fold the excess strip that you trimmed off in half and place it across the two middle squares on the inside of the basket to form a handle. Staple or tape this strip in place.

1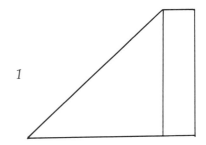

2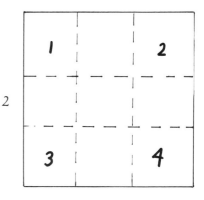

4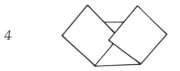

5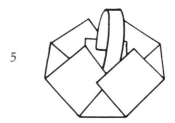

Mother's Day

THIS SPECIAL day began in 1907 when Anna M. Jarvis convinced her church in Philadelphia, Pennsylvania, to hold a special service honoring mothers. As part of this service, everyone wore white carnations. In 1908, a number of Philadelphia churches decided to hold a special mother's day service and agreed on the second Sunday in May for the event. Again, everyone attending wore a white carnation.

By 1911, the observance of the second Sunday in May as Mother's Day had spread throughout the United States. It was also observed that year in churches in Canada, Mexico, South America, Africa, China, and Japan. In 1913, the House of Representatives adopted a resolution stating that all officials of the Federal Government would wear white carnations on the second Sunday in May in honor of Mother's Day. This day was on its way to becoming a national event. Eventually, though, the custom of wearing white carnations was changed to wearing red carnations in honor of a mother who is living and white carnations in honor of a mother who is dead.

This year, why not surprise your mom or someone you'd especially like to remember on this day with a gift that keeps growing. You'll need to start early in April. First, purchase a package of zinnias or other flower seeds. Then collect enough Styrofoam cups to plant five seeds to a cup. Put enough gravel in the bottom of each cup to cover the bottom. Next, fill the cups with potting soil and poke several small drainage holes in the bottom with a sharp pencil. Then plant the seeds following the directions printed on the package. Sprinkle the soil with water and cover each cup with clear wrap. Set the cups on a tray to collect any water that leaks out.

When the sprouts appear in about a week, remove the clear wrap. Move the cup to a warm, sunny spot and continue to water as needed to keep the soil from completely drying out. After the sprouts have grown for a week, remove all but the two biggest plants in each cup. This will allow the plants more growing room. Then, just before you give your growing gift, tie a brightly colored ribbon around each cup.

RIDDLE

What wears a cap but doesn't have a head?

A bottle.

Super Dads

THE THIRD Sunday in June is set aside in the United States to honor human fathers. But there are some animal dads who deserve special recognition, too.

Sea Horse

IT'S COMMON for females to have a special organ in which the young can develop. But this uncommon pop is the one with the brood pouch. When the female sea horse is ready to lay her eggs, she squirts as many as six hundred eggs into the male's pouch. Once the eggs are inside, the pouch seals. Any egg that doesn't nestle into one of the honeycomb compartments in the brood pouch degenerates. Each egg that survives develops a connection with the father's body, receiving oxygen and nutrients directly from him.

After about six weeks, when the young are fully developed, the father goes into labor. Winding his tail around a strand of seaweed as an anchor, he rocks and sways as strong muscular contractions expell his babies one at a time.

Midwife Toad

AFTER THE female midwife toad finishes laying strings of jellylike eggs, the male twists and squirms among them. This causes the egg strands to wrap around his thighs. Then for the next month, the father toad carries the developing young with him. To be sure the eggs don't dry out, this concerned dad even takes time to soak in a puddle. The young tadpoles will need to breathe underwater with gills once they hatch. So when he senses that hatching time is near, the male midwife toad hops to a pond and holds his legs underwater until all the youngsters have hatched.

Greater Rhea

AFTER MATING with several rhea females, the future father rhea tramples the grass flat and scratches out a nesting hole. Then each of his chosen mates lays her eggs in *his* nest—a total of as many as thirty eggs. Weighing as much as eighty pounds and standing nearly five feet tall, this big male bird has no trouble defending his nest. And for the next five weeks, he handles the job of incubating his eggs alone. He won't even allow his mates to come close. The greater rhea male remains a protective father for about four months after the young hatch. He even makes sure the chicks stay dry. If it starts to rain, this dad kneels and spreads his wings over the nest.

Gold-Headed Tamarin

THE MOTHER tamarin gives birth to the babies, but the father takes over their care as soon as they're born. This South American monkey—about the size of a squirrel—leaps from branch to branch among the treetops with two or three babies clinging to him. He cleans them and holds them, handing them back to the mother only when they need to nurse. Later, when the youngsters are old enough to eat solid food, he catches insects and chews them up to produce a monkey baby food. And when they're ready, he teaches the baby monkeys to climb, taking special care that they don't fall.

Parent for a Day

HERE'S A fun way you can experience what it's like to be responsible for a baby. First, hard-boil an egg. After it's cooled, check it over carefully for any cracks. If there are any, mark these with a red ballpoint pen. This egg is going to be your "child" for a day. And as a good egg-parent, your job will be to be sure this egg makes it through the entire day without any new cracks.

Whoa! Before you park your egg baby in a safe place and go off, there is one other rule for egg-parenting. You must remain in contact with your egg at all times. Of course, there are no limits on how you accomplish this. Be creative. Just stay in touch and keep "baby" safe.

For extra fun, challenge your friends to an egg-parenting contest. The rules are the same. And the person who gets their egg through the day with the least cracks is the winner.

At the end of the day, take time to think about what sacrifices you had to make, if any, to be responsible for your charge. Was there any time when the job seemed particularly difficult or boring?

Dyeing Eggs—Naturally

FOR MANY years, the spring equinox was considered the beginning of the new year. Eating eggs was part of sharing in this renewal of life. And eggs dyed bright colors were considered the perfect present for this New Year's festival. Before there were commercial dyes, these gift eggs were colored by boiling them with flowers, leaves, even insect parts. You probably won't want to try bugs as coloring agents, but this year rather than buying packaged dyes, why not use some of nature's colors?

Spinach leaves will give you nice light green eggs. Chop one cup of fresh spinach leaves and add these to about a quart of water in a saucepan. Bring to a boil and let the water continue to boil for about three minutes. Then slip the eggs into the water, remove the pan from the heat, and cover with a lid. Let the eggs sit in the hot water, covered, for twenty minutes. Then slip the eggs out of the water with a large spoon and set them on an oven rack to cool.

Here are some other natural dyes you can try: red cabbage leaves, beets, onion skins, orange peels, and cranberries. You'll need to chop the parts and boil them as you did the spinach. What colors do each of these natural dyes make the eggs? Add more if you want the color to be brighter, but always expect these natural dyes to produce lighter shades than the commercial dyes. If the eggs are uncracked, the natural dyes should not flavor the eggs. But think about your coloring material when you eat one of these eggs. Do spinach-green eggs have a slight spinach taste, for example?

RIDDLE

What goes up white and comes down yellow and white?
·88ə uⱯ

You Crack Me Up

DOES IT seem a shame to throw out the beautifully colored egg shells after you eat the egg? Then you may want to save the shells from a number of different colored eggs and use these to create a mosaic—a picture made up of many tiny individual fragments of color. It's not a new idea. Italian artists were creating egg shell mosaics during the fifteenth century.

To make your egg masterpiece, break the shell fragments into pieces about one-fourth inch across. Sketch the general shape of your design on a sturdy sheet of poster board or cardboard. For best results keep the design simple and divide it into sections—one for blue, green, and so forth—like puzzle pieces. Next, coat the section you plan to work on with white glue and arrange the shell pieces on this glue base. When the mosaic is complete and the glue is dry, you'll probably want to spray it with a protective coat of varnish.

Egg Treasures

THE MOST spectacular Easter eggs ever produced were probably the jeweled eggs created by Peter Carl Fabergé for the Russian tsars Alexander III and Nicholas II. These eggs were particularly special because, in addition to the elaborately decorated exterior, each held a surprise.

The very first Fabergé Easter egg was a gold egg enameled white to look like an ordinary hen's egg. When it was opened, it revealed a gold yolk. Inside the yolk was a chicken made of several different shades of gold. Inside the chicken was a model of the imperial crown, and inside the crown, there was a tiny ruby egg. The tsar was so pleased that Fabergé was given a standing order to produce one egg each Easter with the only rule being that it must contain a surprise.

Fabergé kept his bargain and each egg he made was uniquely beautiful and unusual. The 1908 egg, for example, was made of rock crystal and set on a silver pedestal. Inside, there was a golden tree adorned with jeweled flowers. An enameled peacock sitting in the tree was a windup toy that would strut, spread its tail, and move its head when placed on a table.

The End (for This Year)

ONCE THE tree leaves have all opened, spring is over. The days will be hot and the heat, like the sunlight, will linger late into the evening. Spring flowers will fade, leaving developing seeds and fruit. Eggs will hatch, and babies born during the spring will grow bigger. Green plants will be silently using sunlight to manufacture food. And everywhere you look, you'll see animals busily collecting food—munching on it and storing it—for themselves and for their families.

There will be a lot of new things for you to do, but spring exploring—that fresh, creative time of discovery—is over.

You'll have to wait until the next time it's spring to investigate and find those seasonal surprises again.

INDEX

Allergy, 67–68
American alligator, 33
Anglerfish, 108
Animal disguises and
  tricks, 107–109
Antelope, 32

Baby-sitting business,
  58–60
Baltimore oriole, 39
Bat, 65
Bird
  Baby development
    inside egg, 46–47
  Courtship, 27–32
  Eggs, 43–51
  Molting, 19
  Nest building, 37–40
  Nest soup, 40
  Territorialism, 32
Bison, 32
Bloodroot, 70
Bowerbirds, 27–28
Bubbles, 11–12
Bumblebee moth, 109
Butterflies, 43, 109

Cleaning, 9–10
Chicken egg, 44–49
Clouds
  As weather forecasters,
    101–103
Constellations
  Hercules, 8–9
  Leo, 8–9
Courtship, 27–32

Deer, 35, 55
Dolphin, 57
Dragonfly, 108
Duck, 19

Eagle, 31, 38

Egg
  Chick development
    inside, 46–47
  Dyeing with natural
    dyes, 117–118
  Experiments with,
    47–48
  Fabergé Easter eggs,
    119–120
  Hard-boiled snack, 48
  Parts of, 44–47
  Production, 43–44
  Races, 49
  Shell mosaic, 118–119
  Shell structure, 47
Elephant, 56–58

Fathers (animal), 113–116
Frog, 43, 55, 56

Games
  Bubble challenges, 12
  Egg races, 49
  Hopscotch, 20–22
  Marbles, 22–24
  Smell Hunt, 66
Gold-headed tamarin, 116
Goose, 19
Grass whistle, 74–75
Greater rhea, 115
Grizzly bear, 29

Hepatica (liverwort), 69
Honeybee, 65–66
Hopscotch, 20–22
Horse chestnut, 79
Hummingbird, 28, 39,
  65–66

Inchworm, 108
Io moth, 109

Jack-in-the-pulpit, 74

Kangaroo
  Joey (young), 55
  Milk, 53
Kite
  Building, 94–95
  Flying tips, 93–94
  Flying holiday, 94
  Safety tips, 95
  Why fly, 92–93

Lady's slipper (pink), 73
Landfill, 13–14
Lettuce, 80–82
Liverwort (hepatica), 69

Mantid (flower), 107–108
Maple (sugar), 76, 79–80
Marbles, 22–24
May Day
  Basket, 110–11
  Celebration, 110
Mermaid's purse (shark's
  egg), 51
Midwife toad, 115
Milk
  Animal, 53–54
  Human, 53
  Orange moo drink,
    54–55
Monarch butterfly, 109
Moth, 30, 107, 109
Mother's Day, 112–113
Musk-ox wool (qiviut), 18

Nests, 37–40
Nurseries, 32–39

Opossum, 34–35
Ostrich, 50

Parenting, 116–117
Pheromones, 30–31
Pigeon, 29

Pileated woodpecker, 39
Pinwheel, 89–91
Pollen, 63, 65–68
Pollination, 65–66
Purple martin, 36–37
Pussy willow, 78

Rain gauge, 103–104
Recycling, 14–15
    Paper, 15
Reproduction
    Embryo development, 46–47, 51–52
    Fertilization of egg, 43
    Plant, 63–64
Robin, 6, 32, 37–38

Sea horse, 114
Seal's milk, 53
Shark, 51, 55
Siamese fighting fish, 30

Skunk, 34, 43, 55
Slug caterpillar, 108
Snowshoe rabbit, 28
Soap, 9–12
South American paradoxical frog, 56
Spring
    Equinox, 107
    Seasonal changes, 6–7
    Storm, 97
    Tonic, 15
Strawberry, 16–17
Surface tension, 10
Swan, 55
Swift, 40

Territorialism, 31–32
Thorn bug, 108
Thunder
    Cause, 99
    Toy to simulate, 98–99

Trash, 13–15
Trees
    Blooms, 77–80
    Bud, 77
    Sap, 75–76
Trillium (red), 72
Tulip, 82–84

Umbrella, 99–100
Underwing moth, 107

Viceroy butterfly, 109
Violet (longspur), 71

Weaverbirds, 19
Wildflowers, 68–74
Wind, 65, 87–91
Wind chime, 96–97
Wolf, 57–58

	DATE DUE		
MAR 30			
APR 08			

5206

574.5 MAR Markle, Sandra.

Exploring spring : a season of science activities, puzzles, and games.

FOREST GLEN ELEMENTARY SCHOOL
1935 CARDINAL LN GREEN BAY WI

887194 01186   02249D   19